# Springer Series on Atomic, Optical, and Plasma Physics

Volume 107

The Springer Series on Atomic, Optical, and Plasma Physics covers in a comprehensive manner theory and experiment in the entire field of atoms and molecules and their interaction with electromagnetic radiation. Books in the series provide a rich source of new ideas and techniques with wide applications in fields such as chemistry, materials science, astrophysics, surface science, plasma technology, advanced optics, aeronomy, and engineering. Laser physics is a particular connecting theme that has provided much of the continuing impetus for new developments in the field, such as quantum computation and Bose-Einstein condensation. The purpose of the series is to cover the gap between standard undergraduate textbooks and the research literature with emphasis on the fundamental ideas, methods, techniques, and results in the field.

More information about this series at http://www.springer.com/series/411

Volodymyr Girka · Igor Girka · Manfred Thumm

# Surface Electron Cyclotron Waves in Plasmas

Springer

Volodymyr Girka
School of Physics and Technology
V.N. Karazin Kharkiv National University
Kharkiv, Ukraine

Igor Girka
School of Physics and Technology
V.N. Karazin Kharkiv National University
Kharkiv, Ukraine

Manfred Thumm
Institute for Pulsed Power and Microwave
Technology and Institute of Radio
Frequency Engineering and Electronics
Karlsruhe Institute of Technology
Karlsruhe, Baden-Württemberg, Germany

ISSN 1615-5653 ISSN 2197-6791 (electronic)
Springer Series on Atomic, Optical, and Plasma Physics
ISBN 978-3-030-17114-8 ISBN 978-3-030-17115-5 (eBook)
https://doi.org/10.1007/978-3-030-17115-5



This Springer imprint is published by the registered company Springer Nature Switzerland AG
The registered company address is: Gewerbestrasse 11, 6330 Cham, Switzerland

# Preface

This book has been written as a result of theoretical studies which were carried out during the last forty years. It contains results of comprehensive investigations of properties of surface waves which propagate across an external static magnetic field at harmonics of the electron cyclotron frequency. The main objective of the authors was to collect material devoted to studying different characteristics of these transverse waves, namely—their dispersion properties and their dependence on various design peculiarities of plasma waveguides; their damping due to interaction with the plasma surface (this is the kinetic channel of their power dissipation) and due to collisions between plasma particles (this is the Ohmic channel of their power dissipation); their interaction with flows of charged particles that move above the plasma surface; their parametric excitation due to the effect of an external radiofrequency (RF) field and some aspects of their power transfer for sustaining gas discharges.

Thus, the material collected in this book will be useful for postgraduate plasma physics students and for experts, who are interested in the field of kinetic theory of plasmas or in applications of surface electron cyclotron waves.

The strongest interest of scientists in the development of plasma physics observed during more than the last fifty years is explained by a desire of mankind to use the knowledge gained in the field of research on magnetically confined plasmas for clean energy generation by controlled thermonuclear fusion. During this time, there was a significant evolution in views of physicists on the main mechanism of plasma heating in fusion reactors. Namely, in early years it was planned to utilize Ohmic heating, then at the present time the main mechanism is neutral beam injection (NBI). An additional method of plasma heating that is required for sustaining an optimal heating regime is electron cyclotron resonance heating (ECRH). Improvement of the heating and plasma confining methods in various fusion devices has contributed as well to advances in allied branches of plasma science such as plasma electronics, physics of gas discharges and collective phenomena in solid-state plasmas.

This latter statement is especially valid for cyclotron waves since they are utilized for plasma heating, non-inductive current drive, plasma stabilization and plasma diagnostics in fusion devices and also in all the branches of applied plasma physics mentioned above. However, the features of plasmas which are studied in these fields of applied sciences are different, and the conditions for plasma production and confinement are also different. In addition, it should be mentioned that under laboratory conditions the plasma volume is always finite with boundaries. Restricted space of a plasma especially affects the dispersion properties and spatial field distribution of those waves which can propagate in the plasma.

So we can highlight two joint features of the material presented here. First, magnetoactive plasmas are bounded, and second, the electromagnetic waves which propagate in these plasmas have frequencies at harmonics of the electron cyclotron resonance. This means that all presented theoretical results are obtained in the framework of the kinetic description, and that for all considered problems solutions of the kinetic equation have been obtained using a definite model of interaction between plasma particles and plasma interface.

This monograph contains materials devoted to three types of polarization of surface electron cyclotron waves: TM-modes, X (extraordinary)-modes and O (ordinary)-modes. Unlike the case of propagation of bulk cyclotron waves, surface cyclotron modes are eigenwaves of different bounded plasma structures; therefore, their properties essentially depend on geometrical peculiarities of the studied plasma structure and its design features, and not only on plasma parameters like temperature, particle density and value of an external magnetic field.

The material presented in this monograph is the result of a comprehensive study of these surface cyclotron wave modes. The contents include studying their dispersion properties, calculating their damping rates caused by both the Ohmic and kinetic channel of power dissipation, investigating the interaction between these modes and charged particle beams, studying the parametric excitation of these surface cyclotron modes due to the effect of an external RF field including a non-monochromatic RF field, and mechanisms of power transfer from these surface cyclotron modes to gas discharges.

The start of theoretical studying these eigenwaves has been done by Prof. A. Kondratenko, who was a scientific advisor of V. O. Girka. Later definite support for the investigation of these waves was provided by the Science and Technology Center in Ukraine. Some results of this study of surface electron cyclotron waves have been presented at the Joint Workshops on Electron Cyclotron Emission and Electron Cyclotron Resonance Heating and published in the proceedings of these meetings. Regrettably, V. O. Girka passed away on 3 November 2015, when three of the five chapters of this book had been finished.

I. O. Girka is indebted to the grant of DAAD that allowed him to take part in the final preparation of the text for publication as a book at the Karlsruhe Institute for Technology (KIT). He is also very grateful to Prof. J. Jelonnek and colleagues in the High-Power Microwave Division of the Institute for Pulsed-Power and

Microwave Technology (IHM) at KIT for a lot of fruitful discussions and their kind hospitality. The authors also thank Dr. I. Pavlenko for very fruitful discussions on several issues of this monograph.

Kharkiv, Ukraine Volodymyr Girka
Kharkiv, Ukraine Igor Girka
Karlsruhe, Germany Manfred Thumm

# Contents

# Chapter 1
# Introduction

The term "plasma" for designation of an ensemble of a large number of charged particles was introduced by Tonks and Langmuir [1], who studied electromagnetic waves in low density ionized electrical gas discharges. The statistical properties of such an ensemble of charged particles are determined by the Coulomb interaction between a randomly selected particle and the large number of other surrounding particles, including those which are far away. As a result of the low particle density in the ensemble, this interaction is much stronger than that between adjacent particles. This long-range nature of the Coulomb forces in plasmas gives rise to the possibility for the existence of collective motion of particles.

Often, plasmas are called as "the fourth state of matter". This means that solids, when heated, usually transfer to a liquid state. After further heating they pass into the gaseous state, and at least temperature increasing to values of ~10,000 ÷ 20,000 K leads to significant ionization of neutral gas particles. Such a state of matter is called plasma. The ambient temperature can be lower if there are external sources of ionization and/or recombination is ineffective due to the low plasma density. From the standpoint of classical thermodynamics speaking about plasmas as a new state of matter is not totally correct because the states gas, liquid and solid, which are known as phases, can exist in equilibrium between each other. In contrast to this, under laboratory conditions, significant technological efforts are required to produce and confine plasmas in a discharge chamber or especially in a thermonuclear fusion plasma reactor. The term "equilibrium plasma" is often used to describe the collective motion of charged quasi-particles of any type: conduction electrons, holes, etc. in conductors and semiconductors. If gas plasma particles are characterized by the same temperature, there is a mechanical equilibrium between this plasma and its environment and if there is a balance between energy absorption and emission of the plasma particles, such a plasma is usually considered to be in a quasi-equilibrium state. This means that in such a plasma, there are small fluctuations on the background of a relative equilibrium that is characterized by a Maxwellian distribution function for the plasma particles. The equilibrium can be almost stationary due to the low number of collisions between the plasma particles.



V. Girka et al., *Surface Electron Cyclotron Waves in Plasmas*,
Springer Series on Atomic, Optical, and Plasma Physics 107,
https://doi.org/10.1007/978-3-030-17115-5_1

The circumstance that the plasma volume is restricted significantly affects the dispersion properties and spatial distribution of electromagnetic waves in the plasma. This book describes the results of the research on eigenmodes of electron cyclotron waves propagating just in the peripheral area of bounded plasmas. Therefore, this peculiarity distinguishes this monograph from the long-published basic books that were devoted to description of properties of bulk cyclotron waves propagating in unbounded plasmas [2, 3].

Experimental investigation of plasmas under laboratory conditions is carried out in special devices like magnetic traps and metallic confinement structures. In these confined plasmas, both bulk and surface waves can propagate [4]. Most of the properties of surface and bulk waves are different [5, 6]. Since in various industrial processes just a bounded plasma is produced, very often the conditions just for surface wave propagation are present. These eigenmodes of electromagnetic waves can be effectively excited by various external sources of electromagnetic energy. That is why the properties of electromagnetic eigenwaves propagating in plasma structures are the subject of intensive scientific research.

The properties of surface waves depend in different ways on the type of interfaces on which they propagate. Those surface waves which propagate along plasma-dielectric interfaces are the most studied. The case of surface wave propagation along plasma-metal interfaces are less investigated, although the first experimental results concerning the propagation of high-frequency electromagnetic surface waves along the boundaries of semiconductor plasmas and metals were obtained long time ago [7–9]. Further results on properties of surface waves which propagate along a plane plasma-metal interface are presented in the review [10]. Surface cyclotron X-mode waves were found to propagate along plasma-metal boundaries as well [11, 12].

The results of theoretical studies on electromagnetic phenomena in plasma structures are widely used in plasma [13, 14] and semiconductor [15, 16] electronics. Sometimes, the characteristics of electronic devices with plasma filling are superior to those of vacuum electronics devices. To utilize the electromagnetic power of surface electron cyclotron waves one has to solve the problem how to excite them. One of the most commonly used methods of wave excitation in plasmas is beam-plasma interaction [17, 18]. The theory of this excitation is well developed for the case of unbounded plasmas and the case of plasmas immersed into an infinitely strong magnetic field. However, since the frequencies of electromagnetic waves studied in the present monograph belong to the range of harmonics of the electron cyclotron frequency, magnitudes of an external static magnetic field are considered here as finite. Another commonly used method for excitation of plasma waves is their parametric interaction with external radiofrequency (RF) fields. The basic theory of parametric excitation of plasma waves is described in the monograph [19] for the case of unbounded plasmas. However, the theory of parametric instabilities for surface waves is developed very weakly. It has been therefore interesting to investigate parametric excitation of surface electron cyclotron waves. The results of these studies for three different types of wave polarization are presented in this monograph. In addition, the influence of a non-monochromatic external RF field on parametric instabilities of these modes is considered as well.

Recently, interest to study the properties of electromagnetic surface waves had grown mostly due to their utilization in modern plasma technologies such as pumping of lasers, producing special coating layers (with the required properties), treatment of interfaces of solids, including polishing and etching, growing nanostructures in plasma discharges, designing plasma sources and so on. Development and practical utilization of plasma sources, which apply an external magnetic field, is one of the major directions in the development of plasma technology in many countries [20, 21]. In this case application of surface waves for sustaining gas discharges is characterized by the following significant advantages compared to other types of discharges: the efficiency of excitation of surface waves by external sources of energy is very high, since these waves are eigenmodes of discharge chambers; the absorption of surface plasma waves is higher than that of bulk waves, and therefore their ionization ability is also higher; the repeatability and stability of discharges sustained by surface waves is better. That is why this book provides basic information about the characteristics of electrodynamic models of gas discharges sustained by surface electron cyclotron waves. Efficiency of different mechanisms of electromagnetic power transfer into the maintained plasma from these waves are studied here as well.

The monograph consists of one further general chapter and in addition three special ones. Chapter 2 contains the results of solving the kinetic Vlasov-Boltzmann equation for the cases of different orientation of an external static magnetic field in respect to the bounded plasma interface. Some sections are devoted to the analysis of the impact of an external RF field on the expressions for the permittivity tensor of bounded magnetoactive plasmas.

Chapter 3 presents the analysis of the dispersion properties of surface electron cyclotron waves of TM polarization. It contains analytical expressions of calculated damping rates of these modes caused by both kinetic and Ohmic dissipation of their energy and the analysis of their ability to interact with flows of charged particles moving over the surface of the plasma. Growth rates of the parametric instability of surface electron cyclotron TM-modes caused by external RF fields are calculated here as well. In addition, the possibility to control the scenario of this instability by application of an additional RF field with other values of operating frequency and amplitude is analyzed in this chapter. Mechanisms of power transfer by these TM-modes and their ability to sustain a gas discharge are also discussed here.

Chapter 4 is devoted to results of studies on the characteristics of surface electron cyclotron X-modes. Their dispersion relations are determined here. It is shown that they can propagate along the boundaries of both plasma-vacuum and plasma-metal interfaces. The mechanisms of their damping are discussed and the damping rates caused by kinetic and Ohmic dissipation of their energy are calculated here. It is shown that these X-modes can be excited due to beam-plasma instability. Their growth rates under the regimes of resonant beam and dissipative instabilities are calculated for the model of charged particle flow moving over the plasma surface. The theory of parametric excitation of surface electron cyclotron X-modes, including the analysis of the impact of a non-monochromatic external RF field on the scenario of this instability, is presented here. The final section is devoted to the electrodynamic model of gas discharges which are sustained by these X-modes.

Chapter 5 contains the results of the investigation of the properties of surface electron cyclotron O-modes. In the approach of weak plasma spatial dispersion, the dependence of their frequency on the plasma parameters and the wave vector is derived. Their damping rates caused by interaction of plasma particles with the plasma interface (kinetic damping) and by collisions between plasma particles (Ohmic damping) are calculated here. The interaction between a flow of charged particles moving over the plasma interface and surface electron cyclotron O-modes is discussed and growth rates of these O-mode instabilities are calculated. The theory of their parametric instabilities which take place due to the influence of an external RF field has been derived for the two cases of monochromatic and non-monochromatic RF fields. Kinetic and collisional mechanisms of this O-mode power transfer into plasmas are considered and the volume of the maintained plasma is estimated with the goal to utilize gas discharges for industrial applications.

We have decided to use cgs-units for the formulation of equations since they are more convenient for theoretical description in electromagnetic theory than SI-units. However, all numerical example data will be presented in SI-units.

## References

1. Tonks, L., & Langmuir, I. (1929). Oscillations in ionized gases. *Physical Review, 33,* 195–211.
2. Akhiezer, A. I., Akhiezer, I. A., Polovin, R. V., Sitenko, A. G., & Stepanov, K. N. (1975). *Plasma electrodynamics*. Oxford: Pergamon Press.
3. Lominandze, D. G. (1981). *Cyclotron waves in plasma*. Oxford: Pergamon Press.
4. Landau, L. D., & Lifshits, Y. M. (1960). Course of theoretical physics. In *Electrodynamics of continuous media* (Vol. 8). Oxford: Pergamon Press.
5. Kondratenko, A. N. (1976). *Plasma waveguides*. Moscow: Atomizdat. (in Russian).
6. Kondratenko, A. N. (1985). *Surface and bulk waves in bounded plasma*. Moscow: Energoatomizdat. (in Russian).
7. Toda, M. (1964). Propagation of waves in a solid state plasma waveguide in a transverse magnetic field. *Journal of the Physical Society of Japan, 19*(7), 1126–1130.
8. Hirota, R. (1964). Theory of a solid state plasma waveguide in transverse magnetic field. *Journal of the Physical Society of Japan, 19*(7), 1130–1134.
9. Hirota, R., & Suzuki, K. (1966). Propagation of waves in a bounded solid state plasma in transverse magnetic field. *Journal of the Physical Society of Japan, 21*(6), 1112–1118.
10. Azarenkov, N. A., & Ostrikov, K. (1999). Surface magnetoplasma waves at the interface between a plasma-like medium and a metal in a Voigt geometry. *Physics Reports, 308,* 333–428.
11. Azarenkov, N. A., Girka, V. A., & Sporov, A. E. (1997). Ion cyclotron surface waves on the plasma-metal interface. *Plasma Physics Reports, 23*(3), 231–236.
12. Azarenkov, N. A., Girka, V. O., Kondratenko, A. M., et al. (1997). Surface waves on the harmonics of the electron cyclotron frequency propagating along a plasma-metal interface. *Plasma Physics and Controlled Fusion, 39,* 375–388.
13. Kuzelev, M. V., & Rukhadze, A. A. (1990). *Electrodynamics of dense electron beams in plasmas*. Moscow: Nauka. (in Russian).
14. Kondratenko, A. N., & Kuklin, V. M. (1988). *Principles of plasma electronics*. Moscow: Energoatomizdat. (in Russian).
15. Beletskiy, N. N., Bulgakov, A. A., Khankina, S. I., & Yakovenko, V. M. (1984). *Plasma instabilities and nonlinear phenomena in semiconductors*. Kyiv: Naukova Dumka. (in Russian).

16. Bass, F. G., Bulgakov, A. A., & Tetervov, A. P. (1989). *High frequency properties of semiconductors with superlattice*. Moscow: Nauka. (in Russian).
17. Miller, R. B. (1982). *An introduction to the physics of intense charged particle beams*. New-York: Springer.
18. Humphries, S. (1990). *Charged particle beams*. New-York: Wiley Inc.
19. Silin, V. P. (1973). *Parametric influence of high power emission on plasmas*. Moscow: Nauka. (in Russian).
20. Ferreira, C. M., & Moisan, M. (1993). Microwave discharges: Fundamentals and applications. *NATO Advanced Study Institute, Series B: Physics, 302,* 187–544.
21. Aliev, Yu. M., Schlüter, H., & Shivarova, A. (2000). *Guided-wave produced plasmas*. New-York: Springer.

# Chapter 2
# Methods of Solving the Kinetic Vlasov-Boltzmann Equation in Case of Bounded Magnetized Plasmas

There are two commonly used theoretical methods to study plasma properties, namely, the fluid and kinetic descriptions. The first one is most popular because it is more simple from the mathematical point of view and due to a more evident method of theoretical investigation. Unlike that, the kinetic method is more complicated from the mathematical point of view and more difficult for understanding. However, there are some important phenomena in plasmas which can be adequately described and explained only in the framework of the kinetic description. Among them are: damping and excitation of plasma waves in anisotropic plasmas, propagation of cyclotron waves etc.

This monograph is devoted just to investigation of various properties of electron cyclotron waves in bounded plasmas. They can be analyzed only employing the kinetic approach. Therefore, it is suitable to start the present considerations from studying methods of solving the kinetic equation.

## 2.1 Peculiarities of Fluid and Kinetic Descriptions of Electromagnetic Waves in Plasmas

Most often the model of continuous medium that conducts electric current is applied for an approximate description of plasma properties. That approach, which theoretically describes the properties of conducting gases and liquids using the joint solution of hydrodynamical equations and the Maxwell equations, is called the magnetohydrodynamic approach [1, 2]. The equation of motion of a conducting medium has the form of Newton's second law:

$$\frac{\mathrm{d}\vec{\upsilon}}{\mathrm{d}t} = -\frac{1}{\rho}\mathrm{grad}(P) + \frac{\left[\vec{j}, \vec{H}\right]}{c\rho}, \tag{2.1}$$



V. Girka et al., *Surface Electron Cyclotron Waves in Plasmas*, Springer Series on Atomic, Optical, and Plasma Physics 107, https://doi.org/10.1007/978-3-030-17115-5_2

where $\rho$ is the mass density of the medium, $P$ is its thermodynamical pressure, $\overrightarrow{H}$ is the vector of the magnetic field, $\vec{j}$ is the electric current density, $\mathrm{d}\vec{\upsilon}/\mathrm{d}t$ is the acceleration, that characterizes the motion of an elementary plasma volume caused by the force of thermodynamical pressure and the ponderomotive force which is proportional to $\left[\vec{j}, \overrightarrow{H}\right]$, the vector product of $\vec{j}$ and $\overrightarrow{H}$. Note that (2.1) is an ideal equation in the sense that it lacks any dissipative forces.

A special feature of the magnetohydrodynamic approach applied for the description of the motion of a plasma volume is that this volume contains a large number of charged plasma particles. This method of studying the motion of the plasma particles as motion of a continuum, when a trajectory of motion of the selected elementary volume of the given medium is traced analytically, is called as a description in Lagrangian coordinates [3]. The acceleration in this motion determined for the selected elementary volume is equal to the total time derivative of its velocity $\mathrm{d}\vec{\upsilon}/\mathrm{d}t$, that is calculated along its trajectory. In studying the motion of the plasma particles as the motion of a continuum in the coordinate space, Euler coordinates are often used as well. In this case studies on plasma particle motion are carried out in some selected point of the coordinate space, and the change of the velocity of the plasma elementary volume $\partial\vec{\upsilon}/\partial t$ is calculated just there. Therefore, there is no doubt that the Euler time derivative of the velocity is an acceleration of plasma particles from the physical point of view. The relationship between Lagrangian and Eulerian time derivatives is as follows:

$$\mathrm{d}\vec{\upsilon}/\mathrm{d}t = \partial\vec{\upsilon}/\partial t + (\vec{\upsilon}, \nabla)\vec{\upsilon}. \tag{2.2}$$

Herein and in the following the symbol $\nabla$ is used for the Hamiltonian vector differential operator. For example, in the case of a Cartesian coordinate system it reads as follows: $\nabla = \vec{e}_x\partial/\partial x + \vec{e}_y\partial/\partial y + \vec{e}_z\partial/\partial z$. The symbol $\left(\vec{a}, \vec{b}\right)$ denotes the inner product of the vectors $\vec{a}$ and $\vec{b}$.

The magnetohydrodynamic approach is often complemented by the "cold plasma" condition, that is described mathematically by the following inequality:

$$\beta_{\mathrm{pl}} = 8\pi\ P/H^2 \ll 1. \tag{2.3}$$

Meeting the condition (2.3) means that studies of plasma electromagnetic oscillations are carried out with almost complete neglecting the thermal motion of the plasma particles and taking into account the average plasma motion caused by external forces. This leads to the fact that in the framework of magnetohydrodynamics, it is impossible to adequately account for the finite radius $\rho_L$ of the cyclotron orbit of charged plasma particles ($\rho_L$ is called the Larmor radius) in an external magnetic field, wherein the plasma is placed. There is also no possibility to take into account the ability of plasma particles to gyrate strictly at harmonics of the cyclotron frequency. Therefore, in this case the electric current density is determined as follows:

$$\vec{j} = \sum_{\alpha} e_\alpha n_\alpha \vec{v}_\alpha. \tag{2.4}$$

In (2.4), $e_\alpha$ and $n_\alpha$ are the charge and the density of plasma particles. The summation index indicates the type of charged plasma particles, and the velocity $\vec{v}_\alpha$ is the same (average value) for all particles of one type. Consequently, this approach cannot describe those plasma waves which are connected with the fact that the charged plasma particles can gyrate along the Larmor orbits not only with an angular velocity that is equal to the cyclotron frequency of particles of the corresponding type, but also at the higher harmonics of the cyclotron frequency (electron or ion). Thus for describing electromagnetic waves at harmonics of the cyclotron frequency, which is the subject of this monograph, the magnetohydrodynamic approach is not suitable.

This problem can be solved only within the kinetic approach. This allows one to properly take into account the finite size of the Larmor radius of rotation of charged plasma particles in the external static magnetic field and the influence of chaotic thermal motion on the velocity of plasma particles in the selected elementary plasma volume. To achieve this, one can use the concept of the distribution function of particles over their velocities $f_\alpha(\vec{v})$, that determines the number of particles located in a selected unit plasma volume with velocities in a certain interval. When speaking about the number of plasma particles with a definite velocity one means a time-averaged value of this number. This approach is very fruitful for a variety of problems when studying the properties of many-particle systems in physics. Thus, the description of the state of plasma particles with the use of the distribution function is of statistical nature. Therefore, $f_\alpha(\vec{v})$ is the function that is obtained as a result of averaging the fluctuations which take place due to the particle chaotic thermal motion. The number of particles d$n$ in the unit plasma volume with the velocities which belong to the three-dimensional velocity interval $\mathrm{d}\vec{v}$ (it is also often called as elementary volume in the phase space of velocities) is determined as follows: $\mathrm{d}n = f_\alpha(\vec{v})\mathrm{d}\vec{v}$. It should be emphasized once more that $\mathrm{d}\vec{v}$ here is not a differential of the velocity vector. Its dimensionality is $(\mathrm{m/s})^3$ and in Cartesian coordinates it is equal to the product $\mathrm{d}\vec{v} = \mathrm{d}v_x \mathrm{d}v_y \mathrm{d}v_z$. In the kinetic approach, in contrast to formula (2.4) written for the case of the magnetohydrodynamic approach, the electric current density of particles of type $\alpha$ is determined in an integral way:

$$\vec{j} = \int e_\alpha f_\alpha(\vec{v})\vec{v}\,\mathrm{d}\vec{v}. \tag{2.5}$$

The change of the distribution function $f_\alpha(\vec{v})$ of a certain type of plasma particles with respect to time and space is described by the kinetic Boltzmann equation. This equation can be derived with the use of the Liouville theorem. The physical meaning of this theorem consists in the statement of conservation of the size of the phase volume of a large ensemble of particles (time invariance) for an arbitrary motion without collisions [4]. Since the value of the phase volume does not change with time, the Liouville theorem leads to the conclusion that the value of the plasma particle density in the phase space for an ensemble with a large constant number of particles is preserved during time. Hence, the total time derivative of the distribution function $f_\alpha(\vec{\upsilon})$, that is calculated along the trajectory of the particles, is equal to zero: $\mathrm{d}\, f_\alpha/\mathrm{d}\, t = 0$. The expression for the total time derivative of the distribution function in terms of partial derivatives allows one to derive kinetic equations for plasma particles in the absence of collisions:

$$\frac{\partial f_\alpha}{\partial t} + \sum_{j=1}^{3} \frac{\partial f_\alpha}{\partial x_j} \frac{\partial x_j}{\partial t} + \sum_{j=1}^{3} \frac{\partial f_\alpha}{\partial \upsilon_j} \frac{\partial \upsilon_j}{\partial t} = 0. \tag{2.6}$$

In (2.6) the symbols of summation indicate that derivatives can be calculated over all three coordinates in the coordinate space and over all three components of velocity in the phase space of velocities.

It should be noted that the presence of collisions between the plasma particles can be described by the integral of collisions $(\partial f_\alpha/\partial t)_{\mathrm{col}}$, which was set to zero up to now, but in the general case should be entered on the right side of (2.6). This monograph is restricted either to the case of plasmas without collisions or to the case of the "relaxation time approximation". The latter means the case of few collisions when the effective collision frequency is much lower than the frequency of the studied electromagnetic waves ($\nu_\alpha \ll \omega$). In the first of these cases, the kinetic Boltzmann equation is valid in the form (2.6). In the second case, one can add on the right hand side of (2.6) an approximate expression for the collision integral, whose value is proportional to $\nu_\alpha$.

Since the derivative over time, which results from the Liouville theorem that is formulated for a constant value of the phase volume of a large ensemble of plasma particles, is calculated along the trajectory of the plasma particle, the derivatives $\partial x_j/\partial t = \upsilon_j$ are just the components of the velocity and the derivatives $\partial \upsilon_j/\partial t = a_j$ are the components of the acceleration of plasma particles, respectively. Since the acceleration of a particle with a certain mass $m_\alpha$ is determined by the force $\overrightarrow{F}$ acting on the particle, under the condition that the analytical expression for this force is known and that it does not depend on the distribution function, the kinetic equation (2.6) can be, in principle, integrated. The kinetic equation (2.6) can be written in the form of the kinetic Boltzmann equation:

$$\frac{\partial f_\alpha}{\partial t} + (\vec{\upsilon}, \nabla) f_\alpha + \left( \overrightarrow{F}, \frac{\partial}{\partial \vec{\upsilon}} \right) f_\alpha/m_\alpha = 0. \tag{2.7}$$

One should keep in mind that the real dynamics of these particles depends on their interaction between each particle with all the other particles. In fact, the transition from the phase space of velocities, where the kinetic equations (2.7) are determined, to the canonical phase space (in which every state of a particle is described by generalized coordinates and generalized momenta), can transform (2.7) into the Liouville equation [4]. In the absence of interaction between the particles this allows one to apply the transition procedure from the equation for the distribution function describing an ensemble of $N$ particles to the equation for the distribution function describing an ensemble of $N - 1$ particles, and so on, until obtaining the equation that describes the behavior of a single particle. However, due to the fact that the elementary plasma volume contains a very large number of particles, one can implicitly conclude that it is absolutely impossible to obtain the mathematically correct solution of the Liouville equation for the case of plasmas.

This mcans that one should find and apply an appropriate approximate method in order to solve this problem. In the case of theoretical research in plasma physics, often the self-consistent field approximation is employed [1, 2, 4]. In this approximation, the force $\overrightarrow{F}$ acting on the plasma particles is calculated on the basis of an averaged value of the distribution function, ignoring the plasma fluctuations. Then, after the force $\overrightarrow{F}$ is known, one can calculate the plasma distribution function that produces the force field, which, in its turn, sustains just the calculated distribution function. As an example of a self-consistent field theory one can consider the problem of the Debye electrostatic screening of charged particles in a plasma. In this problem the value of the plasma potential to be substituted into the Poisson equation has just an averaged value.

The self-consistent field approximation is the correct one to describe the long-range forces, because at short distances plasma fluctuations strongly affect the interaction between the plasma particles. Luckily, electromagnetic forces are just long-range forces. That is why the force acting on a plasma particle with charge $e_\alpha$ can be calculated as follows:

$$\vec{F} = e_\alpha\left(\vec{E} + \left[\vec{v} \times \vec{H}\right]/c\right). \tag{2.8}$$

In (2.8), $\overrightarrow{E}$ and $\overrightarrow{H}$ are the self-consistent values of the electric and magnetic fields. They are obtained from solving the Maxwell equations in which the charge density $\rho$ and electric current density $\vec{j}$ are calculated with the help of an averaged distribution function:

$$\rho = \sum_\alpha e_\alpha \int f_\alpha(\vec{v})\mathrm{d}\vec{v}, \tag{2.9}$$

$$\vec{j} = \sum_\alpha e_\alpha \int f_\alpha(\vec{v})\vec{v}\,\mathrm{d}\vec{v}. \tag{2.10}$$

Here the index $\alpha$ indicates the different types of charged particles in the plasma. After one has calculated self-consistently the force $\overrightarrow{F}$ from (2.8) for the plasma particle with mass $m_\alpha$, and substituted the ratio $\overrightarrow{F}/m_\alpha$ into the kinetic Boltzmann equation (2.7), this equation transforms into the kinetic Vlasov-Boltzmann equation:

$$\frac{\partial f_\alpha}{\partial t} + (\vec{v}, \nabla) f_\alpha + \frac{e_\alpha}{m_\alpha}\left\{\left(\overrightarrow{E} + \frac{1}{c}\left[\vec{v}, \overrightarrow{H}\right]\right) \cdot \frac{\partial}{\partial \vec{v}}\right\} f_\alpha = 0. \tag{2.11}$$

The inaccuracy of the self-consistent field approximation is associated with the fact that one neglects fluctuations while calculating the interaction forces between the plasma particles. This is necessary to make it possible to calculate the force $\vec{F}$ using the averaged distribution function. This approximation is valid in the linear approach with respect to the amplitudes of the electromagnetic fields. Thus, the self-consistent field method is the most suitable method for a linear kinetic theory of plasma physics. In this linear approach the self-consistent field approximation to study the surface electron cyclotron waves is applied.

The term "cyclotron waves" is used to call those electromagnetic waves [1, 2] in plasmas which propagate strictly across an external static magnetic field (or at an angle that is close to $\pi/2$). Their frequencies are close to the harmonics $n\omega_\alpha$ of the cyclotron frequency ($n$ is a positive integer; $\omega_\alpha = e_\alpha B_0/(cm_\alpha)$ is the cyclotron frequency of $\alpha$ type particles) in the limits of either very large ($k\rho_L \to \infty$) or very small ($k\rho_L \to 0$) wavenumbers. It is convenient to compare the Larmor radius with the wavelength, because $\rho_L$ is the natural scale describing the gyration of charged particles in a magnetic field.

Since application of plasma theory methods to study the collective motion of conduction electrons in metals or other media with essential magnetic properties is out of the scope of the present considerations, the magnetic permeability in the plasma is assumed to be equal to unity with high accuracy. The difference between the values of magnetic induction $\vec{B}$ and magnetic field strength $\vec{H}$ in the plasma is neglected. Therefore, we apply here the symbol $\vec{B}$ to denote external fields and the symbol $\vec{H}$ for eigenfields of electromagnetic waves in the plasma. As already mentioned above, the case of purely transverse cyclotron waves when the wave vector is directed only across the external magnetic field is under consideration. For those readers who are interested in the case of quasi-perpendicular propagation of cyclotron waves (the angle between the wave vector and the external magnetic field is almost but not equal to $\pi/2$) the [1, 5] are recommended.

It should be emphasized that surface waves, in contrast to the bulk waves, are essentially two-dimensional objects [6, 7]. That is why when choosing the orientation of the coordinate system for studying surface waves it is impossible to use only one component in the wave vector of the investigated electromagnetic wave. This is explained by the fact that surface waves propagate along the surface of the plasma (thus one can apply at least one coordinate along the interface), and they spatially attenuate in the direction perpendicular to the plasma interface (thus one can choose the second coordinate being perpendicular to the interface). Moreover, the component of the wave vector that is perpendicular to the plasma boundary cannot have an arbitrary value [6], its value is determined by the properties of the plasma and the considered wave via application of appropriate boundary conditions. This component has either a complex or a purely imaginary value. The reciprocal value of the imaginary part of the transverse (with respect to the plasma interface) component of its wave vector is called as 1/e-penetration depth of the wave field into the plasma. The penetration depth determines how far from the plasma interface the wave power decreases by an order.

To find the components of the plasma conductivity tensor $\sigma_{ij}\left(\omega, \vec{k}\right)$ and the related plasma permittivity tensor $\varepsilon_{ij}$:

$$\varepsilon_{ij}\left(\omega, \vec{k}\right) = \delta_{ij} + 4\pi i \sigma_{ij}/\omega, \tag{2.12}$$

one has to linearize the kinetic Vlasov-Boltzmann equation (2.11). This means that one assumes that the distribution function $f_\alpha$ consists of two terms: the equilibrium (averaged) distribution function $f_{0\alpha}$ and a small correction $f_{1\alpha}$, $f_\alpha = f_{0\alpha} + f_{1\alpha}$, here $f_{1\alpha} \ll f_{0\alpha}$. In (2.12), $\omega$ and $\vec{k}$ are the angular frequency and wave vector of the studied electromagnetic disturbance, and $\delta_{ij}$ is the Kronecker symbol. For the equilibrium distribution function $f_{0\alpha}$ the two first terms in formula (2.11) vanish due to Liouville's theorem, and only that part in the brackets of the third term remains, which is connected with the external magnetic field because there are no fluctuations of the electromagnetic field in the equilibrium state. Therefore, one obtains the following equation for the distribution function $f_{0\alpha}$:

$$\frac{e_\alpha}{m_\alpha c}\left[\vec{\upsilon}, \vec{B}_0\right] \cdot \frac{\partial}{\partial \vec{\upsilon}} f_{0\alpha} \equiv \omega_\alpha \frac{\partial f_{0\alpha}}{\partial \varphi} = 0, \tag{2.13}$$

here $\varphi$ is the azimuthal angle in the phase space of velocities. By the way, it should be noted here that any collisional integral $(\partial f_\alpha/\partial t)_{\text{col}}$ on the right hand side of (2.11) would not change (2.13), because this integral is equal to zero in the zeroth approximation. Thus, the equilibrium part of the distribution function is an axially symmetric function which only depends on the transverse and parallel (with respect to the induction vector $\vec{B}_0$) components of the plasma particle velocities, $f_{0\alpha} = f_{0\alpha}\left(\upsilon_\perp, \upsilon_\parallel\right)$, and it does not depend on the azimuthal angle $\varphi$. Thus, the equilibrium distribution function is isotropic along the azimuthal direction, however, the question remains about its dependence on the velocities $\upsilon_\perp$ and $\upsilon_\parallel$. If this dependence would be anisotropic, i.e. if there would be an asymmetry with respect to the longitudinal velocity, this would result in the presence of a static electric current in the plasma. If the current in the plasma is absent, then the equilibrium distribution function in the case of a homogeneous background should be of the form $f_{0\alpha} = f_{0\alpha}\left(\upsilon_\perp^2, \upsilon_\parallel^2\right)$. But to find a solution of the kinetic equation it is necessary to calculate the derivative of $f_{0\alpha}$ over velocities, thus one should know the specific form of the equilibrium function $f_{0\alpha}$. Anisotropic distribution functions are usually represented analytically as the product of several Maxwellian distribution functions at different temperatures. In the case of a low-density plasma the expression of $f_{0\alpha}$ in general depends on the initial conditions of plasma production. If the plasma is sufficiently dense, the mean free path of the plasma particles is relatively small, so that this plasma quickly establishes a state of thermodynamic equilibrium, which corresponds to the Maxwell distribution function. This can be written as follows:

$$f_{0\alpha} = \frac{n_{0\alpha}\exp\left(-\upsilon^2/\upsilon_{T\alpha}^2\right)}{\pi^{3/2}\upsilon_{T\alpha}^3}, \tag{2.14}$$

where $\upsilon_{T\alpha} = \sqrt{2T/m_\alpha}$ is the thermal velocity of the plasma particles, and $n_{0\alpha}$ is the equilibrium value of the plasma density.

For the first correction to the distribution function (or perturbed distribution function) one can write down the following equation:

$$\frac{\partial f_{1\alpha}}{\partial t} + (\vec{\upsilon}, \nabla) f_{1\alpha} + \frac{e_\alpha}{m_\alpha}\left\{\left(\vec{E} + \frac{1}{c}\left[\vec{\upsilon}, \vec{B}_0\right]\right) \cdot \frac{\partial}{\partial \vec{\upsilon}}\right\} f_{0\alpha} - \omega_\alpha \frac{\partial f_{1\alpha}}{\partial \varphi} = \left(\frac{\partial f_{1\alpha}}{\partial t}\right)_{\text{col}}, \tag{2.15}$$

where the right hand side of the kinetic Vlasov-Boltzmann equation contains the derivative $(\partial f_{1\alpha}/\partial t)_{\text{col}}$, which is the integral of collisions. Its value in the relaxation time approximation is $(\partial f_{1\alpha}/\partial t)_{\text{col}} = -\nu_\alpha f_{1\alpha}$, and in the limiting case of the absence of plasma particle collisions it is equal to zero. Integration of the linearized kinetic equation (2.15) allows one to find the perturbed distribution function $f_{1\alpha}$ as a weighted linear combination of the components of the electric field of the studied electromagnetic waves: $f_{1\alpha} = \sum_j \left(\partial f_\alpha / \partial E_j\right) E_j$. This expression is convenient for introducing the tensor of plasma conductivity $\sigma_{ik}$ into the kinetic theory of plasma oscillations. Indeed, if external electric currents are absent in the plasma, one can just insert the perturbed distribution function $f_{1\alpha}$ into formula (2.10) as the distribution function of plasma particles, i.e.:

$$j_n = \sum_\alpha q_\alpha \sum_j E_j \int \frac{\partial f_\alpha}{\partial E_j} \upsilon_n \mathrm{d}\vec{\upsilon}. \tag{2.16}$$

Comparing formula (2.16) with Ohm's law in differential form $j_{ik} = \sigma_{ik} E_k$, one can determine the plasma conductivity tensor as a solution of the kinetic Vlasov-Boltzmann equation:

$$\sigma_{nj} = \sum_\alpha q_\alpha \int \upsilon_n \frac{\partial f_\alpha}{\partial E_j} \mathrm{d}\vec{\upsilon}. \tag{2.17}$$

Inserting (2.17) into (2.12) one can calculate the components of the plasma permittivity tensor. It should be noted again that these tensor properties are gained from the model of self-consistent field, so it is suitable for describing only linear perturbations of the electromagnetic field in an unbounded plasma. For the case of a bounded plasma, to find a solution of the kinetic equation (2.15) it is necessary to apply a particular model of the interaction of the plasma with its boundary [8].

Note also that to derive a dispersion relation for electromagnetic waves propagating in plasmas the permittivity tensor $\varepsilon_{ik}$ can be used. Application of the tensor $\varepsilon_{ik}$ is convenient to write the Maxwell equations and to formulate boundary conditions for electromagnetic fields in the plasma. Specific expressions for the tensor components $\varepsilon_{ik}$ for different cases (various orientations of the external magnetic field in respect to the plasma surface, presence of an RF field, etc.) will be obtained in the following sections.

## 2.2 Calculation of Dielectric Permittivity Tensor for Semibounded Gyrotropic Plasmas

A plasma that occupies a region of space $z > 0$ is considered (see Fig. 2.1). It is bounded by a dielectric in the plane $z = 0$. The permittivity of the dielectric is $\varepsilon_d$. An external static magnetic field $\overrightarrow{B}_0$ is oriented perpendicularly to the surface of the plasma, along the $\vec{z}$ axis. Such a plasma, that is immersed into a static external magnetic field oriented perpendicularly to the plasma interface, is often called as a gyrotropic plasma [5, 8]. This is due to the fact that the motion of plasma particles along the normal to the surface is the same as if the magnetic field would not be present, therefore, certain properties of gyrotropic plasmas are similar to those of free plasmas.

The presence of the external magnetic field complicates the oscillation spectrum of a plasma compared with the case of a free plasma. New branches of electromagnetic waves, which in principle cannot exist in free plasmas, arise. For example, cyclotron waves can propagate in magnetoactive plasmas. The hydrodynamic theory, although it is simple and intuitive, is not suitable for the description of the properties of cyclotron waves. The fluid theory is not able to adequately describe plasma phenomena, if the mean free path (or the distance that a particle passes in the plasma during the period of wave oscillation) is more than the skin depth, or if the plasma boundary changes the spatial distribution of reflected particles, or if one wants to correctly describe the phenomena caused by the finite value of the Larmor radius of the plasma particles, etc. In all these cases, it is necessary to use the kinetic theory [7].

The kinetic theory of bounded plasmas is more complicated as compared with the case of unbounded plasmas. This is due to the necessity of modeling the plasma interface in order to be able to write down mathematically the boundary conditions

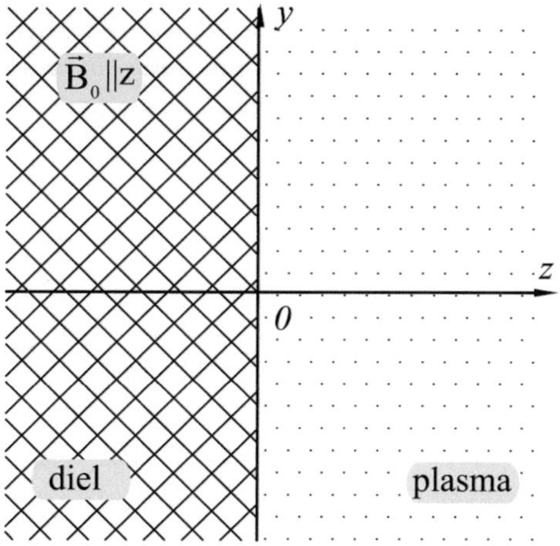


**Fig. 2.1** Schematic of the problem "Semibounded Plasma-Dielectric"

for the distribution function of plasma particles. To somewhat simplify the task of describing the plasma confined in a restricted volume, the plasma interface is usually considered to be quite sharp. In this case, the thickness of the plasma transition layer is considered to be less than the characteristic scale of the problem. For magnetoactive plasmas, this scale is the average value of the Larmor radius [8].

The boundary conditions for the distribution function depend on the type of surface that limits the plasma, or on the method of plasma gas sustaining. In the case of solid-state plasmas, the boundary conditions depend on the method of surface treatment. For gas plasmas, which are confined by a magnetic field, reflection of plasma particles from the plasma boundary is usually considered to be close to elastic and is therefore described by the model of mirror interaction. If the distribution function of particles, which interact with the plasma surface, is an equilibrium one, which means that these particles have lost their ordered speed or a gradient of the spatial distribution of these particles is absent, then such reflection is called to be diffusive. This type of interaction between plasma particles and plasma boundary is characteristic for solid-state plasmas, metals and semiconductors. However, a special treatment of the surface of solid-state plasmas can also make the interaction between plasma particles and its interface close to be a mirror reflection. These types of interactions mentioned above correspond to two limiting cases. In the case of mirror reflection, the distribution function of plasma particles that are incident on the surface remains unchanged after reflection, whereas in the case of diffusive reflection the distribution function maximally varies, namely it turns to the equilibrium state. The real interaction between plasma particles and its surface, of course, is more complex. It has features of both diffusive and mirror reflections [8]. Therefore, from the standpoint of theoretical studying the propagation of electromagnetic waves in plasmas, the following questions are important: how easy is the process of solving the kinetic equation; what is the influence of the choice of the interaction model describing the collisions between plasma particles and the plasma surface; to which extent the results obtained in these limiting cases are different.

The expression, which describes the boundary condition for the distribution function of different particles, can be written in general form. This expression has the same form for ions and electrons if the character of collisions with the plasma boundary for charged plasma particles with opposite charges is assumed to be similar:

$$f_{1\alpha}\left(z_{\text{bound}}, \upsilon_{x,y}, \upsilon_z > 0\right) = p f_{1\alpha}\left(z_{\text{bound}}, \upsilon_{x,y}, \upsilon_z < 0\right). \tag{2.18}$$

In expression (2.18), the parameter $p$ determines the model of interaction between plasma particles and its interface: $0 \le p \le 1$ (with $p = 1$ for mirror reflection, and $p = 0$ for diffuse reflection); $z_{\text{bound}}$ is the normal coordinate which defines the boundary of the plasma.

Another condition for the perturbed distribution function concerns its periodicity along the azimuthal angle, $f_{1\alpha}(\varphi) = f_{1\alpha}(\varphi + 2\pi)$. This means that the gyration of plasma particles around the magnetic field force lines $\left(\vec{B}_0 \| \vec{z}\right)$ must be periodic. The perturbed distribution function in this case can be chosen in the following way:

$$f_{1\alpha} = f_{1\alpha}(z)\exp[i(k_1 x - \omega t)]. \tag{2.19}$$

Then, taking into account that $\partial f_{0\alpha}/\partial\vec{\upsilon} = -2\vec{\upsilon}f_{0\alpha}/\upsilon_{T\alpha}^2$ for a Maxwellian distribution function, and that the cross product $\left[\vec{\upsilon}, \vec{B}_0\right]$ is perpendicular to the velocity $\vec{\upsilon}$, one can simplify the kinetic equation (2.15):

$$\upsilon_z\frac{\partial f_{1\alpha}}{\partial z} - if_{1\alpha}\omega_1 + \omega_\alpha\left(\upsilon_x\frac{\partial f_{1\alpha}}{\partial \upsilon_y} - \upsilon_x\frac{\partial f_{1\alpha}}{\partial \upsilon_y}\right) = -e_\alpha\left(\vec{\upsilon}, \vec{E}\right)\frac{\partial f_{0\alpha}}{\partial \varepsilon_\alpha}. \tag{2.20}$$

In (2.20), $\omega_1 = \omega + i\nu_\alpha$, $e_\alpha$ is the charge of the particles of type $\alpha$, $\varepsilon_\alpha = m_\alpha\upsilon^2/2$ is their kinetic energy, and $\omega_\alpha = e_\alpha B_0/(m_\alpha c)$ is their cyclotron frequency. The solution of (2.20) can be presented as the following sum:

$$f_{1\alpha}(z, \vec{\upsilon}) = \upsilon_x f_\alpha^{(1)}(z, \upsilon_z) + \upsilon_y f_\alpha^{(2)}(z, \upsilon_z). \tag{2.21}$$

This form allows one to introduce the following expressions for the electric current density and electric field: $j^{(\pm)} = j_x \pm ij_y$; $E^{(\pm)} = E_x \pm iE_y$; and as well for the first correction term of the distribution function: $f_\alpha^{(\pm)} = f_\alpha^{(1)} \pm if_\alpha^{(2)}$. Then substitution of (2.21) into the kinetic equation (2.20) and setting the coefficients located nearby the velocity components $\upsilon_x$ and $\upsilon_y$ to zero allow one to derive two equations for determination of the auxiliary functions $f_\alpha^{(1,2)}$:

$$\frac{\partial f_\alpha^{(1,2)}}{\partial z} - i\frac{\omega_1}{\upsilon_z}f_\alpha^{(1,2)} - \frac{\omega_\alpha}{\upsilon_z}f_\alpha^{(2,1)} = -\frac{e_\alpha}{\upsilon_z}E_{x,y}\frac{\partial f_{0\alpha}}{\partial \varepsilon_\alpha}. \tag{2.22}$$

The functions $j^{(\pm)}$, which determine the electric current density through these auxiliary functions, are calculated with the help of the following integral expression:

$$j^{(\pm)} = \sum_\alpha e_\alpha \int_{-\infty}^{+\infty} d\upsilon_x \upsilon_x^2 \int_{-\infty}^{+\infty} d\upsilon_y \int_0^{+\infty} d\upsilon_z\left[f_\alpha^{(\pm)}(\upsilon_z) + f_\alpha^{(\pm)}(-\upsilon_z)\right]. \tag{2.23}$$

From the standpoint of general physics (laws of mechanics and electromagnetic theory) in the case of gyrotropic plasmas, the components of plasma particle velocities, which are directed along the magnetic field $\vec{B}_0$, change only their sign, but preserve their magnitude and the components of plasma particle velocities, which are perpendicular to $\vec{B}_0$, do not change at all. This leads to the conclusion that in the case of mirror reflection of the plasma particles from the plasma interface, the distribution function $f_{1\alpha}$ of a gyrotropic plasma does not feel the presence of the boundary oriented perpendicularly to the $\vec{B}_0$ field. That is why the relationship between the Fourier coefficients of the current density and electric field components is the same as in the case of unbounded plasmas:

$$j_i = \sigma_{ik} E_k. \tag{2.24}$$

Indeed, the mirror reflection of plasma particles from the plasma interface in the case of a gyrotropic plasma can be replaced by the motion of particles from outside of the plasma in the case absence of this boundary.

In the monograph [8], the expression for the additional term of the electrical conductivity tensor of a magnetized unbounded plasma is presented. It is valid in the case of another model of plasma particle reflections when the parameter $p < 1$. Then for a gyrotropic plasma layer with thickness $a_{pl}$, the Fourier coefficients of electric current density and electric field are connected in a more complicated manner as compared with (2.24): $j_i = \sigma_{ik} E_k + \sum\limits_{n=0}^{\infty} R(i,n) E_n$. The expression for the $R(m,n)$ is as follows:

$$R(m,n) = (1-p) \sum_{\alpha} \frac{\Omega_\alpha^2 \upsilon_{T\alpha} \gamma_\alpha^2}{2\pi^{3/2} a} \int\limits_0^\infty \frac{z \exp(-z^2)\left[1 + (-1)^{n+m}\right]}{p^2 \exp(i\zeta) - \exp(-i\zeta)} \times \frac{(-1)^m (1-p) + p \exp(i\zeta) - \exp(-i\zeta)}{\left(\gamma_\alpha^2 - \alpha_m^2 \upsilon_{T\alpha}^2 z^2\right)\left(\gamma_\alpha^2 - \alpha_n^2 \upsilon_{T\alpha}^2 z^2\right)} \mathrm{d}z, \tag{2.25}$$

where $\gamma_\alpha = \omega_1 \mp \omega_\alpha$, $\zeta = a_{\mathrm{pl}} \gamma_\alpha / (z \upsilon_{T\alpha})$, $\alpha_m = \pi\, m / a_{\mathrm{pl}}$, and $\Omega_\alpha = \sqrt{4\pi\, e_\alpha^2\, n_{0\alpha} / m_\alpha}$ is the electron plasma (or Langmuir) frequency. Since the value of $R(m,n)$ is proportional to the difference $p - 1$, it is clear that the relationship (2.24) between current density and electric field is the most simple one just for the model of mirror reflection. Keeping in mind this feature and the fact that the additional term $R(m,n)$ affects only the value of a small damping rate of surface cyclotron waves, this problem will be investigated in the next Chap. 3. Here just the case $p = 1$ is investigated.

Under these conditions, the first term in the kinetic equation (2.20) is equal to $ik_3 \upsilon_z f_{1\alpha}$, and the difference located there in parentheses can be rewritten in much more simple form in cylindrical coordinates: $\upsilon_x = \upsilon_\perp \cos\varphi$; $\upsilon_y = \upsilon_\perp \sin\varphi$:

$$\upsilon_x \frac{\partial f_{1\alpha}}{\partial \upsilon_y} - \upsilon_y \frac{\partial f_{1\alpha}}{\partial \upsilon_x} = \frac{\partial f_{1\alpha}}{\partial \upsilon_y}\frac{\partial \upsilon_y}{\partial \varphi} + \frac{\partial f_{1\alpha}}{\partial \upsilon_x}\frac{\partial \upsilon_x}{\partial \varphi} = \frac{\partial f_{1\alpha}}{\partial \varphi}. \tag{2.26}$$

This allows one to write down the kinetic equation (2.20) in a compact form of a first order differential equation for the azimuthal angle:

$$\frac{\partial f_{1\alpha}}{\partial \varphi} = i f_{1\alpha}(a\cos\varphi + b) + \left(\vec{\upsilon}, \vec{E}\right) \frac{\partial f_{0\alpha}}{\partial \varepsilon_\alpha} \frac{e_\alpha}{\omega_\alpha}, \tag{2.27}$$

where $a = k_1 \upsilon_\perp / \omega_\alpha$, $b = (k_3 \upsilon_z - \omega_1)/\omega_\alpha$, and the expression for the scalar product $\left(\vec{\upsilon}, \vec{E}\right)$ should also be written in cylindrical coordinates as $\left(\vec{\upsilon}, \vec{E}\right) = \upsilon_z E_z + \upsilon_\perp \left(E_x \cos\varphi + E_y \sin\varphi\right)$. The general solution of equation (2.27) can be found in a quite simple way:

$$f_{1\alpha} = \frac{e_\alpha}{\omega_\alpha}\frac{\partial f_{0\alpha}}{\partial \varepsilon_\alpha} e^{i(a\sin\varphi + b\varphi)} \left[ \int_0^{\varphi} \left(\vec{v}, \vec{E}\right) \exp[-i(a\sin\varphi_1 + b\varphi_1)]\, d\varphi_1 + C \right]. \tag{2.28}$$

The integration constant $C$ can be derived from the condition of periodicity of $f_{1\alpha}$ with respect to the azimuthal angle, as it was mentioned above. For carrying out the integration over the angle $\varphi_1$ one can apply an expansion of the exponent into a series of Bessel functions $J_s(a)$ of the first kind with integer order $s$ [9]:

$$\exp(-ia\sin\phi) = \sum_{s=-\infty}^{+\infty} J_s(a)\exp(-is\phi), \tag{2.29}$$

while the relation $J_n(z) = (-1)^n J_{-n}(z)$ is valid [9]. It is also convenient to use the following notation for $E^{(\pm)}$ [see explanation to (2.21)], it allows one to replace the trigonometric functions by exponential functions:

$$E_x\cos\varphi + E_y\sin\varphi = \left[E^{(+)}\exp(-i\varphi) + E^{(-)}\exp(+i\varphi)\right]/2. \tag{2.30}$$

Substitution of formulas (2.29) and (2.30) into the integrand of (2.28) allows one to find an explicit expression for the perturbed distribution function:

$$f_{1\alpha} = \frac{ie_\alpha}{\omega_\alpha}\frac{\partial f_{0\alpha}}{\partial \varepsilon_\alpha} e^{ia\sin\varphi} \sum_{s=-\infty}^{+\infty} e^{-is\varphi} J_s(a) \left[ \frac{v_\perp}{2}\left( \frac{E^{(+)}e^{-i\varphi}}{b+s+1} + \frac{E^{(-)}e^{i\varphi}}{b+s-1} \right) + \frac{E_z v_z}{b+s} \right]. \tag{2.31}$$

The expression (2.31) for $f_{1\alpha}$ can be simplified if one takes into account the fact that the sum over the index $s$ has to be calculated for infinite limits. Thus, this value of the index can be changed to any other value characterized by its shift for an arbitrary finite integer number. To simplify the expression (2.31) one can apply also following recurrence relations for Bessel functions [9]:

$$\begin{aligned} J_{n-1}(z) + J_{n+1}(z) &= 2nJ_n(z)/z; \\ J_{n-1}(z) - J_{n+1}(z) &= 2 \cdot \mathrm{d}J_n(z)/\mathrm{d}z. \end{aligned} \tag{2.32}$$

Thus, the final expression for the perturbed distribution function of particles of a gyrotropic semibounded plasma under the conditions of the mirror model for describing the interaction of plasma particles with its interface is as follows:

$$\begin{aligned} f_{1\alpha} = {} & \frac{ie_\alpha}{\omega_\alpha}\frac{\partial f_{0\alpha}}{\partial \varepsilon_\alpha} e^{ia\sin\varphi} \\ & \times \sum_{s=-\infty}^{+\infty} \frac{e^{-is\varphi}}{b+s} \left[ v_\perp\left( \frac{sE_x}{a} J_s(a) + iE_y \frac{\mathrm{d}J_s(a)}{\mathrm{d}a} \right) + E_z v_z J_s(a) \right]. \end{aligned} \tag{2.33}$$

Substitution of the expression for the perturbed distribution function (2.33) into the formula (2.10) for the electric current density allows one to find the value of the perturbed current, which is determined by the presence of a perturbed electric field in the plasma:

$$\vec{j} = \sum_{\alpha} e_{\alpha} \int \frac{ie_{\alpha}}{\omega_{\alpha}} \frac{\partial f_0}{\partial \varepsilon_{\alpha}} e^{ia \sin\varphi}$$

$$\times \sum_{s=-\infty}^{+\infty} \frac{e^{-is\varphi}}{b+s} \left[ \upsilon_{\perp} \left( \frac{sE_x}{a} J_s(a) + iE_y \frac{\mathrm{d}J_s(a)}{\mathrm{d}a} \right) + E_z \upsilon_z J_s(a) \right] \cdot \vec{\upsilon}\,\mathrm{d}\vec{\upsilon}. \quad (2.34)$$

In (2.34), the elementary volume of the phase space of velocities should be written in the cylindrical coordinates, $\mathrm{d}\vec{\upsilon} = \upsilon_{\perp}\mathrm{d}\varphi\,\mathrm{d}\upsilon_{\perp}\,\mathrm{d}\upsilon_z$. This allows one to calculate the integral over the azimuthal angle, using the integral expression for the Bessel functions of the first kind with integer order [9]:

$$J_n(x) = \frac{1}{2\pi} \int_0^{2\pi} \exp[i(x \sin\varphi - n\varphi)]\,\mathrm{d}\varphi. \quad (2.35)$$

At this stage, carrying out this procedure [integration over the azimuthal angle using the formula (2.35)] is possible only for calculating $j_z$, because its integrand has not any dependence on the azimuthal angle. Thus, in this case the product $\vec{\upsilon}\mathrm{d}\vec{\upsilon} = \upsilon_z\upsilon_{\perp}\mathrm{d}\varphi\,\mathrm{d}\upsilon_{\perp}\,\mathrm{d}\upsilon_z$ has no influence on the dependence of the integrand on the azimuthal angle.

Unlike integration in this case, the calculation of the transverse components of the electric current density (in the respect to the orientation of the external magnetic field) is more complex, because in the calculation of $j_x$, for example, the integrand is multiplied by the factor $\cos\varphi$ since $j_x \propto \upsilon_x = \upsilon_{\perp}\cos\varphi$. In a similar way the calculation of the current density $j_y$ is accompanied by the additional appearance of another trigonometric function as a factor of the integrand, namely, $\sin\varphi$. Therefore, these two transverse components of the electric current density should be calculated using another method. At first, it is useful to apply the expansion of an exponential function into a series of Bessel functions of the first kind [see (2.29)]. Thus, integration over the azimuthal angle gives the following expression for the electric conductivity tensor $\sigma_{ij}$ of the plasma, whose general form was derived in the previous section [see (2.17)]:

$$\sigma_{ik}\left(\omega, \vec{k}\right) = \iint \frac{i\Omega_{\alpha}^2}{2\pi\omega_{\alpha}} \sum_{n=-\infty}^{+\infty} R_{ik}(n,a) \cdot \frac{\mathrm{d}f_{0\alpha}/\mathrm{d}(\upsilon^2)}{n+b} \upsilon_{\perp}\mathrm{d}\upsilon_{\perp}\mathrm{d}\upsilon_z. \quad (2.36)$$

In (2.36), the expression for the tensor $R_{ik}(n,a)$ is given by the following matrix:

$$R_{ik} = \begin{pmatrix} v_\perp^2 \frac{n^2}{a^2} J_n^2(a) & i v_\perp^2 \frac{n}{a} J_n(a) J_n'(a) & v_\perp v_z \frac{n}{a} J_n^2(a) \\ -i v_\perp^2 \frac{n}{a} J_n(a) J_n'(a) & v_\perp^2 \left(J_n'(a)\right)^2 & -i v_\perp v_z J_n(a) J_n'(a) \\ v_\perp v_z \frac{n}{a} J_n^2(a) & -i v_\perp v_z J_n(a) J_n'(a) & v_z^2 J_n^2(a) \end{pmatrix}, \quad (2.37)$$

where $J_n'(x) = \mathrm{d}J_n(x)/\mathrm{d}x$. It should be kept in mind that the integration over the transverse velocity component should be performed in the range from 0 to $+\infty$, and over the longitudinal velocity $\mathrm{d}v_z$ from $-\infty$ to $+\infty$.

Integration over $\mathrm{d}v_z$ in the complex plane should be carried out by the rules of calculating contour integrals using Jordan's lemma. Integrals over the longitudinal velocity component are of improper type. This is not connected with the fact that the collision integral is written in the relaxation time approximation. If one takes the effective collision rate to be zero ($\nu_\alpha = 0$) into this approximate expression, it becomes clear that the integrand in expression (2.36) is an infinite sum, each term of which is singular, because in that infinite range of longitudinal velocity values one can always find a value for which the corresponding denominator results in zero. This occurs both under condition of Cherenkov resonance ($n = 0$) [1] and cyclotron resonance as a result of the Doppler effect ($n \neq 0$):

$$k_3 v_z = \omega - n\omega_\alpha. \quad (2.38)$$

By the way, the Cherenkov resonance condition can be realized not for all components of the tensor $R_{ik}(n, a)$, since most of the components are proportional to the index of summation $n$, thus the term for $n = 0$ vanishes. Therefore, the integration of such a resonant denominator results not only into calculation of the integral in its main sense, it also gives rise to half of the residue at a singular point for the longitudinal velocity, where the condition (2.38) is true:

$$\int_{-\infty}^{+\infty} \frac{\mathrm{d}v_z}{v_z k_3 - (\omega - n\omega_\alpha)} = \mathrm{P} \int_{-\infty}^{+\infty} \frac{\mathrm{d}v_z}{v_z k_3 - (\omega - n\omega_\alpha)} + \frac{i\pi}{k_3} \delta\left(v_z - \frac{\omega - n\omega_\alpha}{k_3}\right), \quad (2.39)$$

where $P \int \frac{\mathrm{d}x}{x - x_0}$ is the main value of improper integrals, and $\delta(x - x_0)$ is the Dirac delta function. For a plasma in equilibrium state, such imaginary terms as in (2.39) describe the phenomenon of attenuation, and in the case of a non-equilibrium plasma, their presence in the electrical conductivity tensor means kinetic instability of corresponding electromagnetic waves. Thus, the presence of the summation over index $n$ in the expression (2.36) indicates the expansions of the electromagnetic field perturbations in series over harmonics of the cyclotron frequency, and the physical meaning of the index $n$ is the number of the cyclotron harmonic.

In the following, the attention is focused on the relativistic effects, which are connected with cyclotron rotation of plasma particles. In the kinetic theory, they are relevant to those particles whose thermal velocity according to the Maxwell distribution is close to the velocity of light. According to the relativistic theory, the cyclotron frequency of the particle gyration decreases with increasing velocity:

$$\omega_\alpha^{(\mathrm{relat})} = \omega_\alpha \left(1 - \upsilon^2/c^2\right)^{1/2} \approx \omega_\alpha \left[1 - \upsilon^2/\left(2c^2\right)\right]. \tag{2.40}$$

It is clear that this relativistic effect is of practical importance only for electrons since the ion cyclotron frequency is still small. Taking the relativism of Larmor gyration into account changes the denominator of the integral (2.39):

$$k_3\upsilon_z - \omega + s\omega_\alpha \to k_3\upsilon_z - \omega + s\omega_\alpha\left(1 - \upsilon^2/\left(2c^2\right)\right). \tag{2.41}$$

This shows that the first-order Doppler effect is realized for waves, which have finite values of longitudinal (in the respect to the direction of the external magnetic field $\vec{B}_0$) wavenumber, as it was mentioned above. However, for purely transverse waves ($k_3 = 0$) the second-order Doppler effect (or transverse Doppler effect) takes place through a definite group of plasma particles which has pronounced relativistic properties. We should remind that in the non-relativistic approximation the transverse Doppler effect is absent. Therefore, in the kinetic theory even strictly transverse electromagnetic waves at cyclotron frequency harmonics damp through cyclotron radiation. In astrophysics, this phenomenon is called as synchrotron radiation because the space plasma is characterized by very low density, and therefore the motion of the plasma particles can be described as that of individual charged particles. Therefore, in order to neglect the relativistic effects in the case of studying transverse perturbations of the electromagnetic field in a magnetoactive plasma one should follow the inequality:

$$\frac{\omega - s\omega_\alpha}{\omega} \gg \frac{\upsilon_{T\alpha}^2}{c^2}. \tag{2.42}$$

Now one has to go back to the integral expression (2.36) for the electrical conductivity tensor of the plasma and complete the integration over the transverse and longitudinal velocities. The integration over the transverse velocity component $\upsilon_\perp$ can be done by using the formula for the Weber's second exponential integral [9]:

$$J_{\mathrm{W}} \equiv \int\limits_0^{+\infty} x J_n(\lambda x) J_n(\mu x) e^{-x^2\eta^2/2} \mathrm{d}x = \exp\left[-\left(\lambda^2 + \mu^2\right)/\left(2\eta^2\right)\right] I_n\left(\mu\lambda/\eta^2\right). \tag{2.43}$$

In (2.43) $I_n(z)$ are the modified Bessel functions. They are characterized by the following properties [9].

- The relationship between the modified Bessel functions and Bessel functions of the first kind is as follows: $I_n(z) = i^{-n} J_n(iz)$.

– Changing the sign of the order $n$ does not affect the modified Bessel functions, $I_n(z) = I_{-n}(z)$.

Thus, the integration of the squared Bessel functions of the first kind $J_n(z)$ gives the modified Bessel functions $I_n(z)$. But the matrix (2.37) contains also the square of the derivative of $J_n(z)$ over its argument and the product $J_n(z)J_n'(z)$. For their integration the differential of the Weber formula (2.43) over the index $\lambda$ (or over $\mu$, this does not matter) is used:

$$\frac{\partial J_W}{\partial \lambda} = \frac{\mu I_n'(\lambda\mu/\eta^2) - \lambda I_n(\lambda\mu/\eta^2)}{\eta^2} \exp\left(\frac{\lambda^2 + \mu^2}{-2\eta^2}\right). \tag{2.44}$$

Now the derivative of the formula (2.44) in respect of the other index $\mu$ can be calculated:

$$\begin{aligned}\frac{\partial^2 J_W}{\partial \lambda\, \partial \mu} &= \left[\mu I_n'\left(\frac{\lambda\mu}{\eta^2}\right) + \frac{\lambda\mu}{\eta^2} I_n''\left(\frac{\lambda\mu}{\eta^2}\right) - \frac{\lambda^2 + \mu^2}{\eta^2} I_n'\left(\frac{\lambda\mu}{\eta^2}\right) + \frac{\lambda\mu}{\eta^2} I_n\left(\frac{\lambda\mu}{\eta^2}\right)\right] \\ &\quad \times \exp\left(\frac{\lambda^2 + \mu^2}{-2\eta^2}\right).\end{aligned} \tag{2.45}$$

Since in the matrix (2.37), the arguments of the Bessel functions of the first kind are the same, one can take $\lambda = \mu$ in the further consideration. This allows one to find the following two integral relations, which are associated with Weber's second exponential integral [9]:

$$\begin{aligned}&\int_0^{+\infty} y^2 J_n(\mu\, y) J_n'(\mu\, y) \exp(-\eta^2 y^2/2)\, dy \\ &= \frac{\mu[I_n'(\eta^2/\mu^2) - I_n(\eta^2/\mu^2)]}{\eta^4} \exp\left(\frac{\mu^2}{-\eta^2}\right),\end{aligned} \tag{2.46}$$

$$\begin{aligned}&\int_0^{+\infty} y^3 \left[J_n'(\mu\, y)\right]^2 \exp\left(\frac{\eta^2 y^2}{-2}\right) dy \\ &= \exp\left(\frac{\mu^2}{-\eta^2}\right) \cdot \frac{\mu^2}{\eta^6}\left[I_n''\left(\frac{\eta^2}{\mu^2}\right) + \left(\frac{\eta^2}{\mu^2} - 2\right) I_n'\left(\frac{\eta^2}{\mu^2}\right) + I_n\left(\frac{\eta^2}{\mu^2}\right)\right].\end{aligned} \tag{2.47}$$

The expressions $I_n'(\mu^2/\eta^2)$ and $I_n''(\mu^2/\eta^2)$ are the first and second derivatives of the modified Bessel functions in respect to their argument. One can apply $y_\alpha = k_1^2 v_{T\alpha}^2/(2\omega_\alpha^2)$ as a variable of integration in these Weber integrals.

Integration of the expression (2.36) for the electrical conductivity tensor of a gyrotropic plasma over the longitudinal velocity is complicated because, as it was noted above, it is an improper integral like:

$$W(z) = \int_{-\infty}^{+\infty} \frac{\exp(-\upsilon^2)}{z - \upsilon} \mathrm{d}\upsilon. \qquad (2.48)$$

The analysis of the integral (2.48) shows that it can be rewritten via integrals in which the integrand has no singularity. This can be shown in the following way: one can take the derivative $\mathrm{d}W/\mathrm{d}z$ and try to obtain an auxiliary equation for the function $W(z)$. Then a minus sign will appear and the denominator will have a form that this expression can be integrated by the method of integration by parts. The integral, obtained in this manner, consists of a Poisson integral and a term, which can be expressed as a product of the function $W(z)$ and its argument:

$$\frac{\mathrm{d}W}{\mathrm{d}z} = -2 \int_{-\infty}^{+\infty} \frac{\upsilon \cdot \exp(-\upsilon^2)}{z - \upsilon} \mathrm{d}\upsilon = 2\sqrt{\pi} - 2zW(z). \qquad (2.49)$$

Thus, a linear inhomogeneous differential equation for $W(z)$ is obtained and its solution has the following form:

$$W(z) = \exp(-z^2) \left( 2\sqrt{\pi} \cdot \int_0^z \exp(u^2)\, \mathrm{d}u + C \right). \qquad (2.50)$$

The integration constant $C$ of expression (2.50) is determined from the boundary condition, $W(z = 0) = C$. Putting $z = 0$ in the expression (2.48), one obtains an improper integral with logarithmic singularity at the point $\upsilon = 0$. Its value is equal to half of the residue at this point. This gives $C = -i\pi$ and the improper integral $W(z)$ is given by the following expression [9]:

$$W(z) = \exp(-z^2) \left( 2\sqrt{\pi} \cdot \int_0^z \exp(u^2)\mathrm{d}u - i\pi \right). \qquad (2.51)$$

To calculate the remaining components of the electric conductivity tensor of a gyrotropic plasma one has to integrate the error integral over the longitudinal velocities $\upsilon_z$, what results in two integrals similar to $W(z)$ [9]. They are calculated in the same way. The first of these integrals, which is contained in the expression for the tensor $\sigma_{ik}$ four times (as the function $W(z)$ is present in this tensor), has the form:

$$W_1(z) = \int_{-\infty}^{+\infty} \frac{\upsilon \cdot \exp(-\upsilon^2)}{z - \upsilon} \mathrm{d}\upsilon = zW(z) - \sqrt{\pi}. \qquad (2.52)$$

It is calculated by adding and subtracting the parameter $z$ in the numerator of the integrand. Then this integral breaks into a Poisson integral and an integral similar to the error integral. The second integral expression is only a part of the diagonal element $\sigma_{33}$ of the tensor $\sigma_{ik}$, and it looks like this:

$$W_2(z) = \int_{-\infty}^{+\infty} \frac{\upsilon^2 \cdot \exp(-\upsilon^2)}{z-\upsilon} d\upsilon = z^2 \cdot W(z) - \sqrt{\pi} \cdot z. \tag{2.53}$$

This integral is calculated by adding and subtracting the value of $z^2$ in the numerator. Then one can obtain the product $z^2 W(z)$ from the integral with the numerator $z^2 \exp(-\upsilon^2)$. The other integral allows one to attain two final integrals after the reduction with the denominator. The first of these final integrals is equal to zero, and the second one is the Poisson integral with the factor $z$.

So, after all the procedures of integrating, the electrical conductivity tensor of a gyrotropic semibounded plasma for the case of the mirror reflection model can be presented in the following form:

$$\sigma_{ik}\left(\omega, \vec{k}\right) = \sum_{\alpha} \frac{i\Omega_\alpha^2}{4\pi^{3/2}} \exp(-y_\alpha) \sum_{s=-\infty}^{+\infty} \frac{z_s \cdot B_{ik}}{\omega_1 - s\omega_\alpha}. \tag{2.54}$$

The expression for the matrix $B_{ik}$ is as follows:

$$B_{ik} = \begin{pmatrix} W\frac{s^2}{y_\alpha} I_s(y_\alpha) & iWs\left(I_s' - I_s\right) & W_1\sqrt{\frac{2}{y_\alpha}} s I_s \\ -iWs\left(I_s' - I_s\right) & W\left[\frac{s^2 I_s}{y_\alpha} + 2y_\alpha\left(I_s - I_s'\right)\right] & -iW_1\sqrt{2y_\alpha}\left(I_s' - I_s\right) \\ -W_1\sqrt{\frac{2}{y_\alpha}} s I_s & iW_1\sqrt{2y_\alpha}\left(I_s' - I_s\right) & 2W_2 I_s \end{pmatrix}, \tag{2.55}$$

here $W = W(z)$, $W_1 = W_1(z)$, $W_2 = W_2(z)$, $I_s = I_s(y_\alpha)$, $I_s' = dI_s(y_\alpha)/dy_\alpha$, $y_\alpha = k_1^2 \rho_L^2/2$, and $z_s = (\omega_1 - s\omega_\alpha)/(k_3 \upsilon_{T\alpha})$.

It should be noted that the conductivity tensor of a gyrotropic plasma (2.54) in the kinetic approach contains both Hermitian and anti-Hermitian components regardless which model is chosen for the collision integral. This is true even if one assumes that there are no collisions in the plasma. This fundamentally distinguishes the kinetic from the magnetohydrodynamic approach. Note, that in the limiting case when the waves are propagating exactly across the external magnetic field ($k_3 = 0$), all non-diagonal components of the conductivity tensor $\sigma_{ik}$, which are associated with the motion along the external magnetic field, will vanish because they are proportional to $\upsilon_z$ or to the derivative $\partial f_\alpha / \partial E_z$ [see (2.34)]. Therefore, in this case, all components in the auxiliary tensor $B_{ij}$, which are proportional to the function $W_1(z \to \infty)$ are zero and the shape of the conductivity tensor becomes similar to that one calculated in the case of a cold plasma. However, even in this case, the conductivity tensor $\sigma_{ik}$ of a gyrotropic plasma obtained in the kinetic approach is different from the tensor $\sigma_{ik}$ obtained in the cold plasma approximation due to the fact, that the diagonal components $\sigma_{11}$ and $\sigma_{22}$ calculated in the kinetic approach are not equal to each other.

The general form of the tensor $\sigma_{ik}$ is inconvenient for analytical studies since there is an infinite number of cyclotron harmonics in (2.54). Even if one investigates the case of completely transverse electromagnetic waves ($k_3 = 0$), in the most favorable case one can only replace these sums by certain quadrature formulas that do not allow one to obtain simple analytical expressions in problems of dispersion properties of the studied waves. Therefore, in the general case, one can only rely on numerical calculations. But if one expands the Bessel functions properly in series over the ratio of the Larmor radius and the wavelength of the investigated waves, then for approximate analytical calculations it is sufficient to take into account the smallest terms which do not vanish (i.e. summands with $s = 0; \pm 1 \ldots$) and the term with the number of the cyclotron harmonic in the sums indicated above, for which the corresponding electromagnetic wave is studied.

Another simplification used for analytical studies of wave phenomena in magnetoactive plasmas is the application of the approximations of weak and strong spatial dispersions. The term “spatial dispersion” rises from the dispersion studied in optics, it means the dependence of the electric conductivity tensor on the wave vector, $\sigma_{ik} = \sigma_{ik}\left(\vec{k}\right)$. The type of spatial dispersion is determined by the ratio of the wavelength, or the depth of penetration of the electromagnetic field into the medium, and the mean free path length. The smaller the mean free path length is, the weaker is the spatial dispersion for various optical phenomena.

In formulas (2.54) and (2.55), the component of the wave vector $\vec{k}$ is present in the arguments of the modified Bessel functions $y_\alpha$ and in the arguments of the functions $W(z_s)$ and $W_{1,2}(z_s)$. Thus, the variation of these functions depending on the values of their arguments determines the type of spatial dispersion of magnetoactive plasmas. If the inequality $\left|\vec{k}\vec{\upsilon}\right| \ll |\omega_\alpha|$ holds (here one can consider the averaged value of the thermal velocity of plasma particles as $\vec{\upsilon}$), the components of the plasma conductivity tensor weakly depend on the wavenumbers of the studied electromagnetic wave. Therefore, this case is called as the approximation of weak spatial dispersion. Often the results obtained in this approach are similar to those arising from the hydrodynamic approach, but only the kinetic approach provides a correct account for the finite value of the thermal velocity. The opposite inequality (presented through the Larmor radius), $\left|\vec{k}\right|\rho_L \gg 1$, describes the strong spatial dispersion of a magnetoactive plasma. In this approximation, the kinetic effects have a significant impact on the studied processes of propagation of electromagnetic waves even if the plasma temperature is very low. This situation is realized within the frequency ranges wherein the plasma refractive index has some specific features, such as the range of plasma resonances or nearby the harmonics of the cyclotron frequency.

## 2.3 Calculation of Dielectric Permittivity Tensor for Gyrotropic Plasmas Affected by a Radio Frequency Field Oriented Across an External Static Magnetic Field

This section is devoted to the derivation of the dielectric permittivity tensor of a plasma immersed into an external static magnetic field, which is perpendicular to the plasma interface. Here it is shown, how one can solve the kinetic equation by the method of integration along trajectories for a magnetized plasma that is affected by an external RF field. The kinetic equation of the perturbed distribution function $f_\alpha$ of charged particles for the case of gyrotropic plasmas has the following general form:

$$\frac{\partial f_{1\alpha}}{\partial t}+\vec{\upsilon}\frac{\partial f_{1\alpha}}{\partial \vec{r}}+\vec{F}^{(0)}\frac{\partial f_{1\alpha}}{\partial \vec{\upsilon}}=-\vec{F}^{(1)}\frac{\partial f_{0\alpha}}{\partial \vec{\upsilon}}, \tag{2.56}$$

where $\vec{F}^{(0)}$ is the unperturbed external force, and the force $\vec{F}^{(1)}$ is determined by the perturbed RF fields of the plasma.

In contrast to the kinetic equation (2.15), there is the additional force $e_\alpha \vec{E}_0 \sin(\omega_0 t)$ of the external RF field that affects the plasma particles at the angular frequency $\omega_0$ with the amplitude $\left|e_\alpha \vec{E}_0\right|$. In this case, the equilibrium distribution function $f_{0\alpha}(\vec{\upsilon}, t)$ of plasma particles is given by the Maxwell distribution function $f_M(\vec{\upsilon})$, which takes the RF field into account:

$$f_{0\alpha}(\vec{\upsilon}, t)=f_M\left(\vec{\upsilon}+\frac{e_\alpha \vec{E}_0}{m_\alpha \omega_0}\sin(\omega_0 t)\right). \tag{2.57}$$

Now, the kinetic equation (2.56) can be solved by the method of integration along trajectories of plasma particles, which move in such plasmas. For particles of each type (ions and electrons) one can write the following equations of motion:

$$\begin{cases} \dfrac{\mathrm{d}\vec{r}}{\mathrm{d}t}=\vec{\upsilon}; \\ \dfrac{\mathrm{d}\vec{\upsilon}}{\mathrm{d}t}=\dfrac{e_\alpha}{m_\alpha}\left(\vec{E}+\dfrac{1}{c}\left[\vec{\upsilon}, \vec{H}\right]\right). \end{cases} \tag{2.58}$$

If in (2.58) the electric $\vec{E}=\vec{E}(t)$ and magnetic $\vec{H}=\vec{H}(t)$ fields are determined and initial values of the radius vector $\vec{r}_0$ and velocity $\vec{\upsilon}_0$ are also known, then from a formal point of view, one can find solutions of this set in a general form:

$$\vec{r}=\vec{r}\left(t, \vec{E}, \vec{H}, \vec{r}_0, \vec{\upsilon}_0\right); \quad \vec{\upsilon}=\vec{\upsilon}\left(t, \vec{E}, \vec{H}, \vec{r}_0, \vec{\upsilon}_0\right), \tag{2.59}$$

where the fields $\overrightarrow{E}$ and $\overrightarrow{H}$ are variable parameters and the parameters $\vec{r}_0$ and $\vec{v}_0$ are constants.

Thus, if the motion of the plasma particles is caused only by the influence of the fields $\overrightarrow{E}$ and $\overrightarrow{H}$, that can be determined from Maxwell's equations by the method of self-consistent fields, one can predict the location $(\vec{r};\, \vec{v})$ of a certain elementary volume of the plasma at the next arbitrary time $t$ on the base of its known coordinates $\vec{r}_0$ and $\vec{v}_0$ in phase space at arbitrary time $t_0$. Therefore, if the initial distribution function of the plasma particles $f_\alpha(t_0, \vec{r}_0, \vec{v}_0)$ is known, one can determine the distribution function of particles in general form at arbitrary moment as follows: $f_\alpha(t, \vec{r}, \vec{v}) = f_\alpha(t, \vec{r}, \vec{v}) = f_\alpha\left[\vec{r}_0(t, \vec{r}, \vec{v});\, \vec{v}_0(t, \vec{r}, \vec{v})\right]$. Hence, substituting this solution for the distribution function into the definition of the electric current density $\vec{j}$ presented in the general form (2.10), one can find the specific value of $\vec{j}$ at an arbitrary moment of time. In its turn, substitution of the obtained value of the electric current density into the Maxwell equations allows one to derive a set of integro-differential equations for the electromagnetic fields $\overrightarrow{E}$ and $\overrightarrow{H}$ of the investigated waves. Finally, substitution of the found solution of this set of integro-differential equations into the set (2.58) allows one to close the problem and find its self-consistent final solution with necessary accuracy.

However, of course, the method looks quite simple only at the first glance. The difficulty is that the set of (2.58) has no general analytical solution. This in its turn makes it impossible to find analytical solutions of the above mentioned set of integro-differential equations for the electromagnetic fields. Thus, the implementation of this method faces huge challenges of both principle and practical types. Therefore, in general practice, theoreticians are forced to apply various approximations, which can simplify this complicated procedure, instead of the general method that can be used first of all to solve weakly nonlinear problems.

The proposed method is analogous to the method of characteristics, widely used for solving differential equations with partial derivatives of the hyperbolic type. Trajectories of plasma particles can be just considered as almost the characteristics of kinetic equations. For studying the properties of small oscillations in plasmas the method of trajectories is used to find the perturbed distribution function that satisfies the kinetic equation (2.56). The form of this equation becomes much simpler if one chooses the radius vector $\vec{r}_0$ and the velocity of plasma particle $\vec{v}_0$ at the definite time $t_0$ as new independent variables. These new variables are associated with the conventional variables by the equations of the particles trajectories:

$$\vec{r} = \vec{r}_0 + \int\limits_{t_0}^{t} \vec{v}(t_1)\, \mathrm{d}t_1; \quad \vec{v} = \vec{v}_0 + \int\limits_{t_0}^{t} \frac{\vec{F}}{m_\alpha}(t_1)\, \mathrm{d}t_1, \tag{2.60}$$

where $\overrightarrow{F}$ is the full force (the sum of influences caused by external electromagnetic fields and small internal perturbations of the field) acting on the plasma particle mass. However, these constants $\vec{r}_0$ and $\vec{v}_0$ are not selected in any specific way. For instance, they can be constants of integration of the equation of motion, which are not perturbed by the considered electromagnetic plasma waves. Therefore, for the undisturbed motion described by these variables $\vec{r}^{(0)}$ and $\vec{v}^{(0)}$, the set of equations, which are similar to the system (2.60), can be written in the following way:

$$\vec{r}^{(0)} = \vec{r}_0 + \int_{t_0}^{t} \vec{v}^{(0)}(t_1)\,\mathrm{d}t_1; \quad \vec{v}^{(0)} = \vec{v}_0 + \int_{t_0}^{t} \frac{\vec{F}^{(0)}}{m_\alpha}(t_1)\,\mathrm{d}t_1. \tag{2.61}$$

Now one can find the derivative of the perturbed distribution function over time using the new variables:

$$\left(\frac{\partial f_{1\alpha}}{\partial t}\right)_{\vec{r}_0,\vec{v}_0} = \left(\frac{\partial f_{1\alpha}}{\partial t}\right)_{\vec{r},\vec{v}} + \frac{\partial f_{1\alpha}}{\partial \vec{r}_\alpha}\left(\frac{\partial \vec{r}_\alpha}{\partial t}\right)_{\vec{r}_0,\vec{v}_0} + \frac{\partial f_{1\alpha}}{\partial \vec{v}_\alpha}\left(\frac{\partial \vec{v}_\alpha}{\partial t}\right)_{\vec{r}_0,\vec{v}_0}. \tag{2.62}$$

As it follows from analysis of (2.61), the new variables $\vec{r}_0$ and $\vec{v}_0$ are in zero approximation equal to the characteristics of the unperturbed trajectories of the plasma particles. Therefore, neglecting the small second order corrections for the derivative of the perturbed distribution function $f_{1\alpha}$ over time (2.62) one can write the following approximate expression:

$$\left(\frac{\partial f_{1\alpha}}{\partial t}\right)_{\vec{r}_0,\vec{v}_0} = \left(\frac{\partial f_{1\alpha}}{\partial t}\right)_{\vec{r},\vec{v}} + \vec{v}_\alpha^{(0)}\frac{\partial f_{1\alpha}}{\partial \vec{r}_\alpha} + \frac{\partial f_{1\alpha}}{\partial \vec{v}_\alpha}\frac{\vec{F}^{(0)}}{m_\alpha}. \tag{2.63}$$

The right-hand side of (2.63) is equal to the left-hand side of the linearized kinetic equation (2.56), so it really can be rewritten in a simpler form, using the new variables $\vec{r}_0$ and $\vec{v}_0$, which can be considered as unperturbed radius vector and velocity of the plasma particles during their motion in external electromagnetic fields. Therefore, the kinetic equation (2.56) can be written in the following new form:

$$\left(\frac{\partial f_{1\alpha}}{\partial t}\right)_{\vec{r}_0,\vec{v}_0} = -\frac{\partial f_{0\alpha}}{\partial \vec{v}_\alpha}\frac{\vec{F}^{(1)}}{m_\alpha}. \tag{2.64}$$

Thus for the linearized kinetic Vlasov-Boltzmann equation, the unperturbed trajectories of the motion of plasma particles play the role of characteristics of this differential equation. Since the values of the new variables are not defined in any specific way, the integration of the kinetic equation (2.64) can be presented in the following form:

$$f_{1\alpha}(t,\vec{r}_0,\vec{v}_0) = -\int_{-\infty}^{t} \frac{\vec{F}^{(1)}\big(t_1,\vec{r}^{(0)}(t_1),\vec{v}^{(0)}(t_1)\big)}{m_\alpha}\frac{\partial f_{0\alpha}}{\partial \vec{v}}\mathrm{d}t_1. \tag{2.65}$$

However, one should remind once again, that this integration must be performed along the unperturbed trajectories of the plasma particles, i.e. one should apply the solutions of equations (2.61). If the studied plasma is described by an isotropic equilibrium distribution function, the integrand on the right-hand side of (2.65) can be replaced as follows:

$$\frac{\vec{F}^{(1)}\left(t_1, \vec{r}^{(0)}(t_1), \vec{\upsilon}^{(0)}(t_1)\right)}{m_\alpha}\frac{\partial f_{0\alpha}}{\partial \vec{\upsilon}} = n_{0\alpha} e_\alpha \left(\vec{E}, \vec{\upsilon}^{(0)}\right)\frac{\partial f_{0\alpha}}{\partial \varepsilon_\alpha}. \tag{2.66}$$

Here $\varepsilon_\alpha = m_\alpha \upsilon^2/2$ is the kinetic energy of the plasma particles, like it was in the previous section.

The scalar product of the perturbed electric field in the plasma and the plasma particle velocity can be calculated using the new variables, i.e. for a plane wave $\vec{E}\left(t, \vec{r}^{(0)}, \vec{\upsilon}^{(0)}\right) = \sum\limits_{n=-\infty}^{+\infty} \vec{E}(n) \cdot \exp\left[i\vec{k}\vec{r}^{(0)} - i(\omega + n\omega_0)\, t\right]$. The presence of the Fourier harmonics over the angular frequency $\omega_0$ of the external electric field in the expression for the perturbed electric field is explained by the fact that the plasma is exposed to an external RF field as mentioned at the beginning of this section. Therefore, of course, if there is no external RF field there is no sum over harmonics of the frequency $\omega_0$.

In the following, the method of integration along trajectories is applied to the problem of calculating the electric conductivity tensor of a gyrotropic plasma that is under the influence of an external RF field. To demonstrate its possibilities one considers a semibounded plasma that is immersed into an external static magnetic field $\vec{B}_0 \| \vec{z}$ directed perpendicularly to the plasma surface. The external alternating electric field $\vec{E}_0 \sin(\omega_0 t)$, with an operating frequency $\omega_0$, is assumed to be oriented across the static magnetic field $\vec{B}_0$. Hence, in the present problem the plasma particles are under the action of two forces. The first one $\vec{F}^{(0)} = \frac{e_\alpha}{m_\alpha}\left\{\vec{E}_0 \sin(\omega_0 t) + \frac{1}{c}\left[\vec{\upsilon}, \vec{B}_0\right]\right\}$ is the unperturbed force, which describes the influence of the external RF field on the plasma particles. The second one $\vec{F}^{(1)} = \frac{e_\alpha}{m_\alpha}\left\{\vec{E} + \frac{1}{c}\left[\vec{\upsilon}, \vec{H}\right]\right\}$ is the perturbed value of the force which is caused by the influence of the small disturbances of the electromagnetic eigenfield on the plasma particles.

In this section, an expression for the conductivity tensor $\sigma_{ik}$ of gyrotropic semibounded plasmas in the case of the presence of an external RF field is found. However, the final form is written here only for the four components which will be applied for the further work in the following Chap. 3 devoted to studying the properties of the surface electron cyclotron TM-mode. It should be underlined that the expressions for two of the four components are identical. Because of the symmetry of the problem (as it was discussed in detail in the previous section) it is considered that the electromagnetic waves studied here are characterized by a wave vector that consists of two components: the first one $k_3$ is oriented along the static magnetic field $\vec{B}_0$ and the second one $k_1$ is oriented across it.

The first step in solving this problem is to find the explicit form of the unperturbed velocity of the plasma particles influenced by the external electromagnetic fields. Newton's second law in terms of projections on Cartesian axes is applied. To simplify the calculations one considers the case that the external RF field has only one component, namely $\vec{E}_0 \| \vec{x}$. Then one can write the following set of equations:

$$\begin{cases} \mathrm{d}\upsilon_x^{(0)}/\mathrm{d}t = e_\alpha E_0 \sin(\omega_0 t)/m_\alpha + \omega_\alpha \upsilon_y^{(0)}; \\ \mathrm{d}\upsilon_y^{(0)}/\mathrm{d}t = -\omega_\alpha \upsilon_x^{(0)}; \\ \mathrm{d}\upsilon_z^{(0)}/\mathrm{d}t = 0. \end{cases} \quad (2.67)$$

The third equation of the set (2.67) is the simplest one. One concludes from it that a particle moves uniformly along the static magnetic field, $\upsilon_z^{(0)} = \text{Const}$. The first and second equations form an autonomous subset of equations that is not associated with the third equation. This allows one to conclude that the equilibrium motion along the external magnetic field has no influence on the motion of plasma particles in the transverse direction.

The first differential equation in (2.67) is inhomogeneous. Hence, its total solution consists of the solution of the homogeneous equation that can be obtained from the inhomogeneous one by equating its right-hand side to zero and a partial solution of the inhomogeneous equation. For the homogeneous equations $\mathrm{d}\upsilon_x^{(0)}/\mathrm{d}t = \omega_\alpha \upsilon_y^{(0)}$ and $\mathrm{d}\upsilon_y^{(0)}/\mathrm{d}t = -\omega_\alpha \upsilon_x^{(0)}$ the solutions are as follows:

$$\begin{cases} \upsilon_x^{(0)}(\text{uni}) = C_1 \cos\omega_\alpha t + C_2 \sin\omega_\alpha t = \upsilon_\perp^{(0)} \cos(\omega_\alpha t - \phi_0), \\ \upsilon_y^{(0)}(\text{uni}) = -C_1 \sin\omega_\alpha t + C_2 \cos\omega_\alpha t = -\upsilon_\perp^{(0)} \sin(\omega_\alpha t - \phi_0), \end{cases} \quad (2.68)$$

where the velocity $\upsilon_\perp^{(0)}$, the starting phase $\varphi_0$, and $C_{1,2}$ are the constants of integration in Cartesian and cylindrical coordinate systems. The transition in the set of equations (2.68) from Cartesian to the cylindrical coordinates is connected with the fact that in the further solving of the investigated problem one needs to integrate over velocities by calculating the density of electric currents in the gyrotropic semibounded plasma just in a cylindrical coordinate system.

To find the partial solution of the first equation of the set (2.67) that describes the dynamics of the motion of the plasma particles one can apply the operation of finding its time derivative and replace there the derivative $\mathrm{d}\upsilon_y^{(0)}/\mathrm{d}t$ by its expression from the second equation of set (2.67):

$$\mathrm{d}^2\upsilon_x^{(0)}/\mathrm{d}t^2 + \omega_\alpha^2 \upsilon_x^{(0)} = e_\alpha E_0 \omega_0 \cos(\omega_0 t)/m_\alpha. \quad (2.69)$$

Substituting of $\upsilon_x^{(0)}$ into (2.69) in a form that resembles the right-hand side of this equation, one can find the expression for the requested partial solution:

$$\upsilon_x^{(0)}(\text{nonuni}) = \frac{e_\alpha \omega_0 E_0 \cos(\omega_0 t)}{(\omega_\alpha^2 - \omega_0^2) m_\alpha}. \quad (2.70)$$

Using the expressions of the solutions of the homogeneous equation (2.68) and inhomogeneous equation (2.70) one obtains a first order differential equation for the equilibrium value of the coordinates $x^{(0)}$. Its solution is as follows:

$$x^{(0)} = \frac{e_\alpha E_0 \sin(\omega_0 t)}{\left(\omega_\alpha^2 - \omega_0^2\right) m_\alpha} + \frac{\upsilon_\perp^{(0)}}{\omega_\alpha} \sin(\omega_\alpha t - \varphi_0). \tag{2.71}$$

Since the wave vector of the investigated electromagnetic wave has no component along the $\vec{y}$ axis, the search for an equilibrium value $y^{(0)}$ is not carried out here.

The expression for the first correction to the distribution function of plasma particles in the studied case according to (2.65) using the formulas (2.66), (2.68) and (2.71) is given by:

$$\begin{aligned} f_{1\alpha} = {} & \frac{2e_\alpha f_M}{m_\alpha \upsilon_{T\alpha}^2} \sum_{m=-\infty}^{+\infty} \int_{-\infty}^{t} \mathrm{d}t_1 \Big[ E_3^{(m)} \upsilon_z^{(0)} + E_2^{(m)} \upsilon_\perp^{(0)} \sin(\omega_\alpha t - \varphi_0) \\ & + E_1^{(m)} \cos(\omega_\alpha t - \varphi_0) \Big] \\ & \times \exp\Big[ -i(\omega + m\omega_0) t_1 + i k_3 \upsilon_z^{(0)} t_1 + i a \sin(\omega_\alpha t_1 - \varphi_0) - i a_E \sin(\omega_0 t_1) \Big], \end{aligned} \tag{2.72}$$

where $a = k_1 \upsilon_\perp^{(0)}/\omega_\alpha$, $a_E = k_1 e_\alpha E_0 / \left[\left(\omega_0^2 - \omega_\alpha^2\right) m_\alpha\right]$, and the lower limit of integration (as it was in the previous section) is chosen in such way that $f_{1\alpha}$ should be a function of time and does not depend on the constant of integration. To calculate this complex integral over time one can apply now the formula (2.29), that replaces two exponents in (2.72) by the corresponding Bessel functions:

$$\begin{aligned} f_{1\alpha} = {} & \frac{2e_\alpha f_M}{m_\alpha \upsilon_{T\alpha}^2} \sum_{m=-\infty}^{+\infty} \int_{-\infty}^{t} \mathrm{d}t_1 \Big[ E_3^{(m)} \upsilon_z^{(0)} + E_2^{(m)} \upsilon_\perp^{(0)} \sin(\omega_\alpha t - \varphi_0) \\ & + E_1^{(m)} \cos(\omega_\alpha t - \varphi_0) \Big] \\ & \times \sum_{s=-\infty}^{+\infty} \sum_{p=-\infty}^{+\infty} J_s(a) J_p(a_E) \exp[-i(\omega + m\omega_0 + p\omega_0 \\ & - s\omega_\alpha - k_3 \upsilon_z^{(0)}) t_1 - i s \varphi_0 \big]. \end{aligned} \tag{2.73}$$

In the further considerations in this section, the summation symbols over different Fourier harmonics are not written separately, instead a single summation symbol, for example $\sum_{s,p,m}$, is applied. Nevertheless one can understand that these are three separate operations of summation, which can be calculated independently of each other and in the same range from $-\infty$ to $+\infty$. Then the recurrence formula (2.32) for the Bessel functions of the first kind [9] is used, that allows to remove the harmonic multipliers located nearby the Fourier images $E_{1,2}^{(m)}$ in the integrand and after that to integrate equation (2.73) over time. Thus, one obtains the following expression for $f_{1\alpha}$:

$$f_{1\alpha} = \frac{2e_\alpha f_M}{m_\alpha \upsilon_{T\alpha}^2} \sum_{s,m,p=-\infty}^{+\infty} \left[ E_3^{(m)} \upsilon_z^{(0)} J_s(a) + E_2^{(m)} \upsilon_\perp^{(0)} i \frac{\mathrm{d}J_s(a)}{\mathrm{d}a} + E_1^{(m)} \frac{s\omega_\alpha}{k_1} J_s(a) \right]$$
$$\times J_p(a_E) \frac{\exp\left[-i\left(\omega + (m+p)\omega_0 - k_3\upsilon_z^{(0)} - s\omega_\alpha\right)t - is\varphi_0\right]}{-i\left[\omega + (m+p)\omega_0 - k_3\upsilon_z^{(0)} - s\omega_\alpha\right]}. \tag{2.74}$$

For calculation of the $n$th Fourier harmonic of the electric current density in the plasma, one needs to know the expression for the $n$th Fourier harmonic of the perturbed distribution function (2.74):

$$f_{1\alpha}^{(n)} = \frac{2e_\alpha f_M}{m_\alpha \upsilon_{T\alpha}^2} \sum_{s,m,p=-\infty}^{+\infty} \left[ E_3^{(m)} \upsilon_z^{(0)} + E_2^{(m)} \frac{\mathrm{d}J_s(a)}{\mathrm{d}a} \frac{\upsilon_\perp^{(0)} i}{J_s(a)} + E_1^{(m)} \frac{s\omega_\alpha}{k_1} \right]$$
$$\times \frac{J_s(a) J_{p-m}(a_E) \exp(-is\varphi_0)}{[-i(\omega + (m+p)\omega_0 - k_3\upsilon_z^{(0)} - s\omega_\alpha)]} \int_{-\infty}^{+\infty} \mathrm{d}t \exp[-i(\omega + p\omega_0 - s\omega_\alpha)t]$$
$$\times \exp[i(\omega + n\omega_0)t - ia\sin(\omega_\alpha t - \varphi_0) + ia_E \sin(\omega_0 t)]. \tag{2.75}$$

Now the exponent $\exp(ia_E \sin\omega_0 t)$ in the expression (2.75) can be written using Bessel functions, as it has been already done earlier in this section. Nevertheless the integral over time has not to be calculated at this step of solving the problem yet, because this is associated with the variable angle $\varphi_0$ that is included in the expression for the element of the phase volume of velocities $\mathrm{d}\vec{\upsilon} = \upsilon_\perp^{(0)} \mathrm{d}\varphi_0 \mathrm{d}\upsilon_\perp^{(0)} \mathrm{d}\upsilon_z^{(0)}$. Thus, making these substitutions one obtains the expression for $f_{1\alpha}$ in the final form:

$$f_{1\alpha}^{(n)} = \frac{2ie_\alpha f_M}{m_\alpha \upsilon_{T\alpha}^2} \sum_{s,m,p=-\infty}^{+\infty} \left[ E_3^{(m)} \upsilon_z^{(0)} + E_2^{(m)} \frac{\mathrm{d}J_s(a)}{\mathrm{d}a} \frac{i\upsilon_\perp^{(0)}}{J_s(a)} + E_1^{(m)} \frac{s\omega_\alpha}{k_1} \right]$$
$$\times \sum_{q=-\infty}^{+\infty} \frac{J_s(a) J_{p-m}(a_E) J_{q-n}(a_E) \exp(-is\varphi_0)}{\omega + p\omega_0 - k_3\upsilon_z^{(0)} - s\omega_\alpha}$$
$$\times \int_{-\infty}^{+\infty} \mathrm{d}t \exp[-i(p-q)\omega_0 t + is\omega_\alpha t - ia\sin(\omega_\alpha t - \varphi_0)]. \tag{2.76}$$

Knowing the expression for the perturbed distribution function of plasma particles, one can find the expressions for the Fourier images of the components of the electric current density in the plasma.

One can calculate these expressions using a cylindrical coordinate system. For instance, the integral form of the first component $j_1^{(n)} = \int d\vec{\upsilon}\, e_\alpha f_{1\alpha}^{(n)} \upsilon_x$ of the Fourier image of the electric current density is as follows:

$$j_1^{(n)} = \int \mathrm{d}\vec{\upsilon} \frac{2ie_\alpha^2 f_M}{m_\alpha \upsilon_{T\alpha}^2} \sum_{s,p,m,q} \left[ E_3^{(m)} \upsilon_z^{(0)} + E_2^{(m)} \frac{\upsilon_\perp^{(0)} i}{J_s(a)} \frac{\mathrm{d}J_s(a)}{\mathrm{d}a} + E_1^{(m)} \frac{s\omega_\alpha}{k_1} \right]$$

$$\times \sum_{q=-\infty}^{+\infty} \frac{J_s(a) J_{p-m}(a_E) J_{q-n}(a_E) \exp(-is\varphi_0)}{\omega + p\omega_0 - k_3 \upsilon_z^{(0)} - s\omega_\alpha} \upsilon_\perp^{(0)} \cos(\varphi_0 - \omega_\alpha t)$$

$$\times \int_{-\infty}^{+\infty} dt \exp[-i(p-q)\omega_0 t + is\omega_\alpha t - ia\sin(\omega_\alpha t - \varphi_0)]. \quad (2.77)$$

One can calculate the integral over the azimuthal angle as follows:

$$\int_0^{2\pi} d\varphi_0 \cos(\varphi_0 - \omega_\alpha t) \cdot \exp(ia\sin(\varphi_0 - \omega_\alpha t) - is\varphi_0)$$

$$= \frac{s}{a} \sum_k J_k(a) \exp(-ik\omega_\alpha t) \int_0^{2\pi} \exp[i\varphi_0(k-s)]\mathrm{d}\varphi_0 = \frac{2\pi\, s}{a} J_s(a) \exp(-is\omega_\alpha t). \quad (2.78)$$

Now the integral over time in (2.77) can be calculated by application of the integral expression for the Dirac delta function [9]. This simplifies the form of the product of Bessel functions, whose argument is proportional to the amplitude of the external RF field. The orders of these Bessel functions become more similar to each other: $J_{p-m}(a_E) J_{p-n}(a_E)$. The presence of just this product of Bessel functions distinguishes the expression of the Fourier image of the first component of the electric current density $j_1^{(n)}$ in the case when the plasma is influenced by an external RF field $\vec{E}_0 \sin(\omega_0 t)$ from the appropriate expression obtained in the case that was discussed in the previous section, when the plasma was not influenced by an external electric field $\left(\vec{E}_0 = 0\right)$. This is the only distinguishing feature. One can explain this fact by the following reason. The integration over the longitudinal and transverse velocities in expression (2.77) is to be carried out in the same way as it was demonstrated in the formulas (2.46)–(2.48) and (2.51)–(2.53). Therefore, the integration over the transverse velocity should be carried out using Weber's exponential integral [9] and integrals related to it. Therefore, the procedure of deriving the expressions for the Fourier images of the remaining components of the electric current densities $\left(j_{2,3}^{(n)}\right)$ does not need to be explained in detail. They differ from those obtained in the previous section by only the product of Bessel functions mentioned above, i.e. by $J_{p-m}(a_E) J_{p-n}(a_E)$.

Now the formula for the electric conductivity tensor of a gyrotropic plasma, which is influenced by an external RF field $\vec{E}_0 \sin(\omega_0 t) \| \vec{x}$ oriented along the $\vec{x}$ axis is given by:

$$\sigma_{ik}\left(\omega,\vec{k}\right)=\sum_{\alpha}\frac{i\Omega_{\alpha}^{2}}{4\pi^{3/2}}\exp(-y_{\alpha})\sum_{s,m,p}\frac{z_{s}J_{p-n}(a_{E})J_{p-m}(a_{E})}{\omega_{p}-s\omega_{\alpha}}B_{ik},\tag{2.79}$$

where $\omega_p=\omega+p\omega_0$. The tensor $B_{ik}$ has the same form as in (2.55).

In the next Chap. 3, the properties of surface electron cyclotron TM-modes will be studied under the condition of weak spatial dispersion of gyrotropic plasmas. Therefore, one needs the expressions for the four components of the tensor $\sigma_{ik}\left(\omega,\vec{k},n\right)$ which are necessary for considerations under the approximation of weak spatial dispersion along the normal to the plasma, i.e. when the inequality $z_s=\left(\omega_p-s\omega_\alpha\right)/(k_3 v_{T\alpha})\gg 1$ is true. Under these conditions, the function $W(z_s)$ defined by the integral (2.51) can be written in the following approximate form [9]:

$$W(z_s\gg 1)\approx\frac{\sqrt{\pi}}{z_s}\left(1+\frac{1}{2z_s^2}+\cdots\right)-i\pi\exp\left(-z_s^2\right).\tag{2.80}$$

The explicit expressions for the components of the tensor $\sigma_{ik}\left(\omega,\vec{k},n\right)$ under the approximation of weak spatial dispersion along the normal to the surface of a gyrotropic plasma which will be applied for solving the problems presented in the next Chap. 3 are as follows:

$$\sigma_{11}^{(n)}=\sum_{\alpha}\sum_{s,m,l}\frac{is^{2}\Omega_{\alpha}^{2}\exp(-y_{\alpha})I_{s}(y_{\alpha})}{4\pi\, y_{\alpha}(\omega_{n+m}-s\omega_{\alpha})}J_{m}(a_{E})J_{m-l}(a_{E}),$$

$$\sigma_{13}^{(n)}=\sigma_{31}^{(n)}=\sum_{\alpha}\sum_{s,m,l}\frac{isk_{3}\omega_{\alpha}\Omega_{\alpha}^{2}\exp(-y_{\alpha})I_{s}(y_{\alpha})}{4\pi\, k_{1}(\omega_{n+m}-s\omega_{\alpha})^{2}}J_{m}(a_{E})J_{m-l}(a_{E}),$$

$$\sigma_{33}^{(n)}=\sum_{\alpha}\sum_{s,m,l}\frac{i\Omega_{\alpha}^{2}\exp(-y_{\alpha})I_{s}(y_{\alpha})}{4\pi\,(\omega_{n+m}-s\omega_{\alpha})}J_{m}(a_{E})J_{m-l}(a_{E}),\tag{2.81}$$

where the summation over the index $\alpha$ is carried out for all types of plasma particles, and the summation over the indices $s, m, l$ is done independently from each other in the limits between $-\infty$ and $+\infty$. To simplify the form of the formulas, the limits of summation over the indices $s, m, l$ are not specified. Further, $\omega_{n+m}=\omega+(n+m)\omega_0$, and the other notations are the same as in the previous sections.

## 2.4 Calculation of Dielectric Permittivity Tensor for Gyrotropic Plasmas Affected by a Non-monochromatic Radio Frequency Field Oriented Across an External Static Magnetic Field

Generators of electromagnetic wave energy utilized for plasma heating and excitation of electromagnetic waves in plasmas often provide a non-monochromatic frequency spectrum. Indeed, the frequency spectrum of the waves emitted by these electronic devices (we are not talking about lasers and various quantum generators) is not perfectly narrowband. Besides of that plasmas can be specifically influenced by radiation from several sources with different frequencies of electromagnetic power. This circumstance has determined the content of the present section.

First, one has to derive the expression of the electric conductivity tensor for a gyrotropic plasma influenced by an external non-monochromatic electric field oriented along the surface of the studied semibounded plasma where the electric field is oriented along the $\vec{x}$ axis $\left(\vec{E}_0(t)\|\vec{x}\right)$. The non-monochromatic external RF field is modelled by superposition of two alternating electric fields of different amplitudes and different operating frequencies $\vec{E}_0(t) = \vec{E}_{01}\sin(\omega_{01}t + \beta_1) + \vec{E}_{02}\sin(\omega_{02}t)$. Here, the kinetic Vlasov-Boltzmann equation is solved again by the method of integration along trajectories. To do that, one has to consider at first the equations of motion of the plasma particles. The equations for acceleration along the axes $\vec{y}$ and $\vec{z}$ are the same as in the previous section [see the set of (2.67)]. Therefore, the motion along a stationary magnetic field occurs with a constant velocity:

$$z^{(0)} = \upsilon_z^{(0)}t. \tag{2.82}$$

The equation for the velocity of plasma particles along the axis $\vec{x}$ is given by:

$$\mathrm{d}\upsilon_x^{(0)}/\mathrm{d}t = e_\alpha E_0(t)/m_\alpha + \omega_\alpha \upsilon_y^{(0)}. \tag{2.83}$$

As it was done in the previous section, one can take the derivative of this equation over time and replace $\mathrm{d}\upsilon_y^{(0)}/\mathrm{d}t$ by its expression that is presented in the set (2.67). This allows one to obtain the following differential equation:

$$\mathrm{d}^2\upsilon_x^{(0)}/\mathrm{d}t^2 + \omega_\alpha^2\upsilon_x^{(0)} = \sum_{i=1}^{2} e_\alpha E_{0i}\omega_{0i}\cos(\omega_{0i}t + \beta_i)/m_\alpha, \tag{2.84}$$

where the initial phase $\beta_2 = 0$ for simplicity. Comparing this equation with (2.69), that describes the case of a monochromatic external RF field, one can conclude that replacement of the value $E_0\omega_0\cos(\omega_0 t)$ by the sum of the RF fields $\sum_i E_{0i}\omega_{0i}\cos(\omega_{0i}t + \beta_i)$ makes these equations equal to each other. Hence, the form of solution in the present case is different from the expression (2.71) only by the sum mentioned above:

$$\upsilon_x^{(0)} = \sum_{i=1}^{2} \frac{e_\alpha E_{0i}\omega_{0i}\cos(\omega_{0i}t + \beta_i)}{\left(\omega_\alpha^2 - \omega_{0i}^2\right)m_\alpha} + \upsilon_\perp^{(0)}\cos(\omega_\alpha t - \varphi_0). \quad (2.85)$$

That is why the electric field of the investigated waves changes also weaker as compared with the previous case [see explanation that follows formula (2.66)]. Now it is described by a double sum over harmonics of the two operating frequencies ($\omega_{01}$ and $\omega_{02}$, respectively):

$$\vec{E}\left(t, \vec{r}^{(0)}\right) = \sum_{n_1=-\infty}^{+\infty}\sum_{n_2=-\infty}^{+\infty} \vec{E}(n_1, n_2)\cdot\exp\left[i\vec{k}\cdot\vec{r}^{(0)} - i(\omega + n_1\omega_{01} + n_2\omega_{02})t\right], \quad (2.86)$$

where $\vec{k}\cdot\vec{r}^{(0)} = k_1x^{(0)} + k_3z^{(0)}$ is the scalar product. The expression for $x^{(0)}$ can be obtained by direct integration of (2.85):

$$x^{(0)} = \sum_{i=1}^{2} \frac{e_\alpha E_{0i}\sin(\omega_{0i}t + \beta_i)}{\left(\omega_\alpha^2 - \omega_{0i}^2\right)m_\alpha} + \frac{\upsilon_\perp^{(0)}}{\omega_\alpha}\sin(\omega_\alpha t - \varphi_0). \quad (2.87)$$

Thus, the new (unperturbed) coordinates allow one to present the integral expression for the solution of the kinetic equation in a simple form, which is similar to expression (2.72), obtained in the previous section:

$$\begin{aligned} f_{1\alpha} = {} & \frac{2e_\alpha f_M}{m_\alpha \upsilon_{T\alpha}^2} \sum_{m_1,m_2=-\infty}^{+\infty} \int_{-\infty}^{t} \mathrm{d}t_1\Big[E_3(m_1, m_2)\upsilon_z^{(0)} + E_2(m_1, m_2)\upsilon_\perp^{(0)}\sin(\omega_\alpha t - \phi_0) \\ & + E_2(m_1, m_2)\upsilon_\perp^{(0)}\sin(\omega_\alpha t - \phi_0)\Big]\cdot\exp\big[-i(\omega + m_1\omega_{01} + m_2\omega_{02})t_1 + ik_3\upsilon_z^{(0)}t_1 \\ & + E_2(m_1, m_2)\upsilon_\perp^{(0)}\sin(\omega_\alpha t - \phi_0)\Big]\cdot\exp\big[-i(\omega + m_1\omega_{01} + m_2\omega_{02})t_1 + ik_3\upsilon_z^{(0)}t_1 \\ & + ia\sin(\omega_\alpha t_1 - \phi_0) - ia_{E1}\sin(\omega_{01}t_1 + \beta_1) - ia_{E2}\sin(\omega_{02}t_1)\big] \end{aligned} \quad (2.88)$$

where $a = k_1\upsilon_\perp^{(0)}/\omega_\alpha$, and $a_{E1,E2} = k_1e_\alpha E_{01,02}/\left[\left(\omega_{01,02}^2 - \omega_\alpha^2\right)m_\alpha\right]$ are dimensionless amplitudes of the plasma particle oscillations which occur due to the influence of the two frequency components of the external RF field. In the expression (2.88), summation over the numbers of Fourier harmonics is carried out independently (later in this section, no attention is paid to this issue and the limits of summation for Fourier harmonics are not indicated).

Using the exponential expansion in series for Bessel functions of the first kind (see (2.29)) one can simplify the integrand in the formula (2.88) and obtain the following expression:

$$f_{1\alpha} = \frac{2e_\alpha f_M}{m_\alpha v_{T\alpha}^2} \sum_{m_1,m_2,s} \int_{-\infty}^{t} \mathrm{d}t_1 \Big[ E_3(m_1,m_2) v_z^{(0)} + v_\perp^{(0)} (E_2(m_1,m_2) \sin(\omega_\alpha t - \phi_0)$$

$$+ E_1(m_1,m_2)\cos(\omega_\alpha t - \phi_0))\Big]$$

$$\sum_{p_1,p_2} J_s(a) J_{p1}(a_{E1}) J_{p_2}(a_{E2}) \exp\left[-i\left(\omega - k_3 v_z^{(0)}\right)t_1\right]$$

$$\times \exp[i(s\omega_\alpha - \omega_{02}(m_2+p_2) - \omega_{01}(m_1+p_1))t_1 - ip_1\beta_1 - is\phi_0] \quad (2.89)$$

Now it is demonstrated how to use the recurrent relations for Bessel functions of the first kind [9] to simplify the product $\sin(\omega_\alpha t - \varphi_0)\exp[is(\omega_\alpha t - \varphi_0)]$ located nearby the Fourier image $E_2(m_1, m_2)$ in expression (2.89), because this will facilitate the calculation of the integral over azimuthal angle further:

$$\sum_s J_s(a)\sin(\omega_\alpha t - \varphi_0)\exp[is(\omega_\alpha t - \varphi_0)] = \sum_s i\frac{\mathrm{d}J_s(a)}{\mathrm{d}a}\exp[is(\omega_\alpha t - \varphi_0)]. \quad (2.90)$$

A similar operation can be done in respect of the term present in the integrand of expression (2.89) that is proportional to $\cos(\omega_\alpha t - \varphi_0)$. Thus, after such transformations one obtains only exponential functions, so that the desired integral can be calculated in a relatively easy way:

$$f_{1\alpha} = \frac{2e_\alpha f_M}{m_\alpha v_{T\alpha}^2} \sum_{s,m1,m2} \left[ E_3(m_1,m_2) v_z^{(0)} J_s(a) + E_2(m_1,m_2) v_\perp^{(0)} i \frac{\mathrm{d}J_s(a)}{\mathrm{d}a} \right.$$

$$\left. + E_1^{(m)} \frac{s\omega_\alpha}{k_1} J_s(a) \right] \sum_{p1,p2} J_{p1}(a_{E1}) J_{p2}(a_{E2}) \exp\left(ik_3 v_z^{(0)} t - is\phi_0 - ip_1\phi_0\right)$$

$$\times \frac{\exp[-i(\omega + (m_1+p_1)\omega_{01} + (m_2+p_2)\omega_{02} - s\omega_\alpha)t]}{\left[-i(\omega + (m+p)\omega_0 - k_3 v_z^{(0)} - s\omega_\alpha)\right]} \quad (2.91)$$

Comparing the expression (2.91) with the correspondent expression obtained in the previous section for a monochromatic RF field (2.74), one can conclude that the effect of each component of the external RF field is described by its individual factor in the formula for the perturbed distribution function $f_{1\alpha}$, that is defined only by the characteristics of the individual components of the applied RF field. Thus, this fact simplifies the process of verification of the correctness of the obtained results.

The expression for the Fourier image of the perturbed distribution function $f_{1\alpha}$ reads as follows:

$$
\begin{aligned}
f_{1\alpha}^{(q_1,q_2)} &= \frac{2e_\alpha f_{\mathrm{M}}}{m_\alpha v_{T\alpha}^2} \sum_{s,m1,m2} \left[ E_3(m_1,m_2) v_z^{(0)} + E_2(m_1,m_2) \frac{\mathrm{d}J_s(a)}{\mathrm{d}a} \frac{v_\perp^{(0)} i}{J_s(a)} \right. \\
&\quad \left. + E_1(m_1,m_2) \frac{s\omega_\alpha}{k_1} \right] \\
&\sum_{p1,p2} \frac{J_s(a) J_{p_1}(a_{E1}) J_{p_2}(a_{E2}) \exp(-is\phi_0 - ip_1\beta_1)}{\left[ i\left( s\omega_\alpha - \omega - (m_1+p_1)\omega_{01} - (m_2+p_2)\omega_{02} + k_3 v_z^{(0)} \right) \right]} \\
&\quad + \int\limits_{-\infty}^{+\infty} \mathrm{d}t \exp[-i((p_1+m_1-q_1)\omega_{01} + (p_2+m_2-q_2)\omega_{02} - s\omega_\alpha))t] \\
&\quad \times \exp[i(\omega + n\omega_0)t - ia\sin(\omega_\alpha t - \phi_0) \\
&\quad + ia_{E1}\sin(\omega_{01}t + \beta_1) + ia_{E2}\sin(\omega_{02}t)].
\end{aligned} \tag{2.92}
$$

Now, as it has been done in the previous section, the factor $\exp\left[ia_{Ej}\sin\left(\omega_{0j}t + \beta_j\right)\right]$ is replaced by the corresponding Bessel functions. However, replacement of the summation indices in such a way, that allows one to simplify the expression in the denominator of the integrand, does not make it possible to take the integral over time yet. The reason of that is the same as in the previous section. These two operations allow one to get the final expression for the Fourier image of the perturbed distribution function:

$$
\begin{aligned}
f_{1\alpha}^{(q_1,q_2)} &= \frac{2e_\alpha f_{\mathrm{M}}}{m_\alpha v_{T\alpha}^2} \sum_{s,m1,m2} \left[ E_3(m_1,m_2) v_z^{(0)} + E_2(m_1,m_2) \frac{\mathrm{d}J_s(a)}{\mathrm{d}a} \frac{v_\perp^{(0)} i}{J_s(a)} + E_1(m_1,m_2) \right. \\
&\quad \times \left. \frac{s\omega_\alpha}{k_1} \right] \sum_{p1,p2,n1,n2} \frac{J_s(a) J_{p_1-m_1}(a_{E1}) J_{p_2-m_2}(a_{E2}) J_{n_1-q_1}(a_{E1}) J_{n_2-q_2}(a_{E2})}{\left[ i\left( s\omega_\alpha - \omega - p_1\omega_{01} - p_2\omega_{02} + k_3 v_z^{(0)} \right) \right]} \\
&\quad \times \exp[i((n_1 - q_1 + m_1 - p_1)\beta_1 - s\phi_0)] \\
&\quad \int\limits_{-\infty}^{+\infty} \mathrm{d}t \exp[i((n_1 - p_1)\omega_{01} + (n_2 - p_2)\omega_{02})t] \\
&\quad \times \exp[is\omega_\alpha t - ia\sin(\omega_\alpha t - \phi_0)].
\end{aligned} \tag{2.93}
$$

The main difference of formula (2.93) compared with formula (2.76) is the presence of additional factors in the form of two Bessel functions of the first kind. Their argument is determined by the characteristics of the second component of the two-frequency external RF field. Thus, the process of integration over the transverse and longitudinal velocities in the expressions for the components of the Fourier image of the electric RF current in a gyrotropic plasma that is under the influence of an external RF field that has two components with different amplitudes and operating frequencies, does not depend on both the possible presence or absence and the characteristics of the RF field. Hence, this procedure should be performed in the same way as in the absence of an external RF field, which is described in detail in Sect. 2.2.

However, in order to avoid the repetition of operations which are described there in detail, here only the other components of the Fourier images of the electric current density (let they be $y$th and $z$th components) are calculated without unnecessary refining. Thus, the starting integral expression for $j_2(q_1, q_2)$ can be written as:

$$
\begin{aligned}
j_2(q_1, q_2) = & \int \mathrm{d}\vec{\upsilon} \frac{2ie_\alpha^2 f_M}{m_\alpha \upsilon_{T\alpha}^2} \upsilon_\perp^{(0)} \sum_{s,m1,m2} \Big[ E_3(m_1, m_2) \upsilon_z^{(0)} \\
& + E_2(m_1, m_2) \frac{\upsilon_\perp^{(0)} i}{J_s(a)} \frac{\mathrm{d}J_s(a)}{\mathrm{d}a} + E_1(m_1, m_2) \frac{s\omega_\alpha}{k_1} \Big] \\
& \times \sum_{n1,n2,p1,p2} \frac{J_s(a) J_{p_1-m_1}(a_{E1}) J_{p_2-m_2}(a_{E2}) J_{n_1-q_1}(a_{E1}) J_{n_2-q_2}(a_{E2})}{\omega + (p_1\omega_{01} + p_2\omega_{02}) - k_3\upsilon_z^{(0)} - s\omega_\alpha} \\
& \times \exp(i\beta_1(n_1 + m_1 - p_1 - q_1) - is\varphi_0) \\
& \int_{-\infty}^{+\infty} \mathrm{d}t \sin(\varphi_0 - \omega_\alpha t) \exp[it(s\omega_\alpha - (p_1 - n_1)\omega_{01})] \\
& \times \exp[it(n_2 - p_2)\omega_{02} - ia\sin(\omega_\alpha t - \varphi_0)].
\end{aligned}
\tag{2.94}
$$

In contrast to the expression for $j_1^{(n)}$, where the integral over the azimuthal angle (2.78) has been calculated by replacing $\cos(\varphi)\exp(ia\sin\varphi)d\varphi$ by simply changing the variable $\exp(ia\sin\varphi)d\sin(\varphi)$, here one has to make the replacement $\sin\varphi\, d\varphi = \big[\exp(i\varphi) - \exp(-i\varphi)\big]\mathrm{d}\varphi/(2i)$. Then the exponent function, whose argument is a harmonic function, should be replaced by a series over Bessel functions of the first kind according to (2.29) and the second recurrence relation (2.32) for Bessel functions should be applied. After that, the integral over $\mathrm{d}\varphi_0$ in (2.94) can be calculated easily:

$$
\int_0^{2\pi} \mathrm{d}\varphi_0 \sin(\varphi_0 - \omega_\alpha t) \exp[i(a\sin(\varphi_0 - \omega_\alpha t) - s\varphi_0)] = \frac{-2\pi i}{\exp(is\omega_\alpha t)} \frac{\mathrm{d}J_s(a)}{\mathrm{d}a}. \tag{2.95}
$$

After that, one can take the integral over time in the expression for $j_2(q_1, q_2)$ using the presentation of the Dirac delta function $\delta(z - z_0)$ as a Fourier integral [9]:

$$
\int_{-\infty}^{+\infty} \mathrm{d}t \exp[i\omega_0(m-n)] = 2\pi \cdot \delta(m-n). \tag{2.96}
$$

This procedure reduces the number of summations by a factor of two and therefore allows one to simplify the expression for $j_2(q_1, q_2)$:

$$
\begin{aligned}
j_2(q_1,q_2) = &\int_0^{+\infty} \mathrm{d}\upsilon_\perp^{(0)} \int_{-\infty}^{+\infty} \mathrm{d}\upsilon_z^{(0)} \frac{i\Omega_\alpha^2 (\upsilon_\perp^{(0)})^2}{\pi^{3/2}\upsilon_{T\alpha}^5} \sum_{s,m1,m2,p1,p2} \exp\left[-\upsilon^{(0)^2}/\upsilon_{T\alpha}^2\right] J_s(a) \\
&\times \frac{\mathrm{d}J_s(a)}{\mathrm{d}a}\left[E_3(m_1,m_2)\upsilon_z^{(0)} + E_2(m_1,m_2)\frac{i\upsilon_\perp^{(0)}}{J_s(a)}\frac{\mathrm{d}J_s(a)}{\mathrm{d}a} + E_1(m_1,m_2)\frac{s\omega_\alpha}{k_1}\right] \\
&\times \exp(i\beta_1(m_1-q_1))\frac{J_{p_1-m_1}(a_{E1})J_{p_2-m_2}(a_{E2})J_{p_1-q_1}(a_{E1})J_{p_2-q_2}(a_{E2})}{\omega + (p_1\omega_{01}+p_2\omega_{02}) - k_3\upsilon_z^{(0)} - s\omega_\alpha} \\
&\times \exp[it(s\omega_\alpha - (p_1-n_1)\omega_{01} - (n_2-p_2)\omega_{02}) - ia\sin(\omega_\alpha t - \varphi_0)], \qquad (2.97)
\end{aligned}
$$

where $(\upsilon^{(0)})^2 = \left(\upsilon_\perp^{(0)}\right)^2 + (\upsilon_z^{(0)})^2$. Here the integration over the transverse velocity is not discussed in detail, because the terms which are integrands of the integral (2.97) are proportional to the Fourier images of the electric field components $E_{1,3}(m_1,m_2)$. Therefore they can be calculated applying formula (2.46), since their dependence on $\upsilon_\perp^{(0)}$ looks like $J_s(a)\frac{\mathrm{d}J_s(a)}{\mathrm{d}a}\left(\upsilon_\perp^{(0)}\right)^2$.

Accordingly, the dependence on $\upsilon_\perp^{(0)}$ of the part of the integrand (2.97), which is proportional to $E_2(m_1,m_2)$, is given by $\left(\frac{\mathrm{d}J_s(a)}{\mathrm{d}a}\right)^2 \upsilon_\perp^{(0)^3}$. That is why this part of $j_2(q_1,q_2)$ can be calculated with the aid of formula (2.47) connected with the exponential Weber's integral [9].

Integrals over $\mathrm{d}\upsilon_z^{(0)}$ in the expression (2.97) have been also calculated in Sect. 2.2 through the function $W(z_s)$ and functions, which are connected with it, according to the error integral [9]. Terms, which are proportional to $E_{1,2}(m_1,m_2)$ will be defined through the function $W(z_s)$ and the term, which is proportional to $E_3(m_1,m_2)$, can be expressed after integration through the function $W_1(z_s)$ because its numerator has the multiplier $\upsilon_z^{(0)}$.

An integral expression for the Fourier image $j_3(q_1,q_2)$ can be written in a similar way as the expression (2.94):

$$
\begin{aligned}
j_3(q_1,q_2) = &\int \mathrm{d}\vec{\upsilon}\frac{2ie_\alpha^2 f_M}{m_\alpha \upsilon_{T\alpha}^2}\upsilon_z^{(0)} \sum_{s,m1,m2}\Big[E_3(m_1,m_2)\upsilon_z^{(0)} \\
&+E_2(m_1,m_2)\frac{\upsilon_\perp^{(0)} i}{J_s(a)}\frac{\mathrm{d}J_s(a)}{\mathrm{d}a} + E_1(m_1,m_2)\frac{s\omega_\alpha}{k_1}\Big] \\
&\times \sum_{n1,n2,p1,p2}\frac{J_s(a)J_{p_1-m_1}(a_{E1})J_{p_2-m_2}(a_{E2})J_{n_1-q_1}(a_{E1})J_{n_2-q_2}(a_{E2})}{\omega + (p_1\omega_{01}+p_2\omega_{02}) - k_3\upsilon_z^{(0)} - s\omega_\alpha} \\
&\times \exp(i\beta_1(n_1+m_1-p_1-q_1) - is\varphi_0) \\
&\int_{-\infty}^{+\infty} dt\, \exp[it(s\omega_\alpha - (p_1-n_1)\omega_{01})] \\
&\times \exp[it(n_2-p_2)\omega_{02} - ia\sin(\omega_\alpha t - \varphi_0)]. \qquad (2.98)
\end{aligned}
$$

Here integration over the azimuthal angle can be done easier than in the case of calculating the expression of $j_3(q_1, q_2)$ because the integrand of (2.98) has no harmonic multipliers located nearby an exponential function. Hence, in this case we can use formula (2.35). Integration over time gives the Dirac delta function once again. Thus, as it has been obtained for the expression (2.97), one can derive the following expression for $j_3(q_1, q_2)$:

$$
\begin{aligned}
j_3(q_1, q_2) = & \int\limits_0^{+\infty} \mathrm{d}\upsilon_\perp^{(0)} \int\limits_{-\infty}^{+\infty} d\upsilon_z^{(0)} \frac{i\Omega_\alpha^2 \upsilon_\perp^{(0)} \upsilon_z^{(0)}}{\pi^{3/2} \upsilon_{T\alpha}^5} \sum_{s,m1,m2,p1,p2} \exp\left[-\upsilon^{(0)^2}/\upsilon_{T\alpha}^2\right] J_s^2(a) \\
& \times \left[ E_3(m_1, m_2)\upsilon_z^{(0)} + E_2(m_1, m_2) \frac{i\upsilon_\perp^{(0)}}{J_s(a)} \frac{\mathrm{d}J_s(a)}{\mathrm{d}a} + E_1(m_1, m_2) \frac{s\omega_\alpha}{k_1} \right] \\
& \times \exp(i\beta_1(m_1 - q_1)) \frac{J_{p_1-m_1}(a_{E1}) J_{p_2-m_2}(a_{E2}) J_{p_1-q_1}(a_{E1}) J_{p_2-q_2}(a_{E2})}{\omega + (p_1\omega_{01} + p_2\omega_{02}) - k_3\upsilon_z^{(0)} - s\omega_\alpha}.
\end{aligned}
\tag{2.99}
$$

The integration in expression (2.99) over the perpendicular component of the plasma particle velocity $\mathrm{d}\upsilon_\perp^{(0)}$ can be carried out either with the aid of the second Weber's exponential integral [9] (for the terms which are proportional to the Fourier images $E_{1,2}(m_1, m_2)$), or (this concerns only the term that is proportional to the Fourier image $E_2(m_1, m_2)$) by application of formula (2.46), that is associated with this integral.

Integration over the longitudinal velocity $\mathrm{d}\upsilon_z^{(0)}$ for Fourier images, which are proportional to $E_2(m_1, m_2)$, leads to a result, which can be expressed through the function $W_2(z_s)$. For the term, that is proportional to $E_3(m_1, m_2)$, the result of integration is expressed through the function $z_s W_2(z_s)$. Thus, one can obtain the following final form of the expression for the Fourier image $j_3(q_1, q_2)$:

$$
\begin{aligned}
j_3(q_1, q_2) = & \frac{i\Omega_\alpha^2}{4\pi^{3/2}} \sum_{s,m1,m2,p1,p2} \frac{\exp(-y_\alpha) I_s(y_\alpha)}{\omega_{p1+p2} - s\omega_\alpha} \exp[i\beta_1(m_1 - q_1)] W_2(z_s) \\
& \times \left[ 2E_3(m_1, m_2) z_s + E_2(m_1, m_2) \frac{ik_1\upsilon_{T\alpha}}{\omega_\alpha} \left( \frac{\mathrm{d}I_s(y_\alpha)}{I_s(y_\alpha)\mathrm{d}y_\alpha} - 1 \right) - E_1(m_1, m_2) \frac{s\omega_\alpha}{k_1 k_3} \right] \\
& \times \exp(i\beta_1(m_1 - q_1)) \frac{J_{p_1-m_1}(a_{E1}) J_{p_2-m_2}(a_{E2}) J_{p_1-q_1}(a_{E1}) J_{p_2-q_2}(a_{E2})}{\omega + (p_1\omega_{01} + p_2\omega_{02}) - k_3\upsilon_z^{(0)} - s\omega_\alpha}.
\end{aligned}
\tag{2.100}
$$

Hence, the conductivity tensor of a gyrotropic semibounded plasma in this case is as follows:

$$
\begin{aligned}
\sigma_{ik}\left(\omega, \vec{k}, q_1, q_2\right) = & \sum_\alpha \frac{i\Omega_\alpha^2}{4\pi^{3/2}} \exp(-y_\alpha) \sum_{s,m1,m2,p1,p2} \frac{z_s B_{ik}}{\omega_{p1+p2} - s\omega_\alpha} \\
& \times J_{p1-m1}(a_{E1}) J_{p2-m2}(a_{E2}) J_{p1-q1}(a_{E1}) J_{p2-q2}(a_{E2}),
\end{aligned}
\tag{2.101}
$$

where the tensor $B_{ik}$ has the same form as in (2.55), $\omega_{p1+p2} = \omega + p_1\omega_{01} + p_2\omega_{02}$.

## 2.5 Conclusions

Considering the materials presented here one can conclude that just application of the kinetic approach allows one to describe the collective motion of charged particles in plasmas at harmonics of the electron cyclotron frequency. The reason for this is the ability of the kinetic approach, unlike the hydrodynamic approach, to take into account the finite size of the Larmor radius of the plasma electrons during their rotation in an external static magnetic field. Thus, it is just the electron Larmor radius that is the scale, which can be applied in theoretical studies of plasmas for determination whether the thermal motion of the plasma particles is intensive or not. Therefore, the criterion for the strength of the spatial dispersion of the plasma (in other words—the intensity of the plasma's thermal motion) is the value of the product of electron Larmor radius and wavenumber of the corresponding electromagnetic wave. If this product is larger than one, the spatial dispersion of the plasma is strong. In the opposite case, if this product is less than one the spatial plasma dispersion is weak.

It is shown that there are two commonly used methods of solving the kinetic Vlasov-Boltzmann equation, namely, the method of direct integration over the azimuthal angle and the method of integration along trajectories. In authors' opinion, the second method seems to be more obvious in the case of plasmas irradiated by an external RF field. The influence of an external RF field on the form of the dielectric permittivity tensor of the plasma results in the existence of an additional factor, which is the product of two Bessel functions of the first kind. The argument of these functions is the ratio of electron oscillation amplitude caused by the external RF field and the wavelength of the studied electromagnetic waves.

It is proved that the presence of the plasma boundary leads to the appearance of an anti-Hermitian part of dielectric permittivity tensor of the studied magnetoactive plasma system. Therefore, even in the case of a collisionless plasma, propagation of electromagnetic waves in such a medium will be accompanied by definite wave damping that can be called as kinetic damping. This collisionless damping is analogous to the well-known Landau damping. Damping rates of this kinetic damping are determined mainly by the type of the studied electromagnetic wave and the properties of the related plasma filled metallic confinement system and also by the type of interaction between the plasma particles and the plasma boundary (in much lesser extent).

## References


1. Akhiezer, A. I., Akhiezer, I. A., Polovin, R. V., Sitenko, A. G., & Stepanov, K. N. (1975). *Plasma electrodynamics*. Oxford: Pergamon Press.
2. Krall, N. A., & Trivelpiece A. W. (1986). *Principles of plasma physics*. San Francisco Press.
3. Vucovic, S. (1986). *Surface waves in plasma and solids*. Singapore: World Scientific.
4. Lifshits, Ye. M., & Pitaevsky, L. P. (1981). *Physical kinetics: course of theoretical physics* (Vol. 10). New York: Pergamon Press.

5. Drummond, W. E., & Rosenbluth, M. N. (1962). Anomalous diffusion arising from micro instabilities in a plasma. *Physics of Fluids, 5*(12), 1507–1513.
6. Kondratenko A. N. (1976). *Plasma waveguides*. Moscow: Atomizdat. (in Russian).
7. Landau, L. D., & Lifshits, Ye. M. (1960). *Course of theoretical physics. Electrodynamics of continuous media* (Vol. 8). Oxford: Pergamon Press.
8. Kondratenko A. N. (1979). *Penetration of a field into plasma*. Moscow: Atomizdat. (in Russian).
9. Abramowitz, M., & Stegun I. A. (Eds). (1964). *Handbook of mathematical functions*. New-York: National Bureau of Standards, Applied Mathematics, Series 55.

# Chapter 3
# Surface Electron Cyclotron TM-Mode Waves

Electromagnetic waves in a waveguide filled with magnetoactive plasma are the subject of intense research in the field of plasma physics. A significant part of this research is performed by studying waves at harmonics of ion and electron cyclotron frequencies because cyclotron resonance waves together with Langmuir waves belong to the most interesting phenomena in plasma physics. In addition to this general physical interest, propagation of cyclotron waves and associated processes are widely applicable in various applied problems in the field of plasma physics and gas discharges, plasma electronics and controlled thermonuclear fusion, plasma chemistry and plasma technologies. Up to now the properties of electromagnetic waves at harmonics of the cyclotron frequency of bulk type were studied sufficiently (see monographs [1–3] and references cited therein).

One of the first work that dealt with issues of cyclotron wave propagation in plasmas is the paper of Gross [4]. Bulk cyclotron waves were studied further in [5–9], where it is shown that longitudinal (potential) waves in plasmas at harmonics of the cyclotron frequency can be unstable due to the finite value of the Larmor radius of charged particles gyrating in the external magnetic field. The idea of describing oscillations of conduction electrons in metals as gas of quasi-free particles has allowed one at that time to explain the phenomenon of cyclotron resonance in metals [10], which are under the influence of an external magnetic field that is oriented parallel to their surface. This theory was further developed in [11–13], where certain conceptions have been clarified, a large number of experimental investigations concerning observations of cyclotron resonance in different metals has been explained, the influence of changing the angle of inclination between the external magnetic field and the surface of solids on this resonance has been studied. In addition, it was also used to describe the case of a gas plasma. The study of collective excitations of charge carriers in solids [14–16] allows one to obtain important information about the binding energy, structure of the crystal lattice, the shape of the Fermi surface and other electromagnetic properties of charge carriers in solids.

It is well known that the application of methods, which have been developed in one area of physics, very often can be used also for explanation of phenomena studied in



V. Girka et al., *Surface Electron Cyclotron Waves in Plasmas*,
Springer Series on Atomic, Optical, and Plasma Physics 107,
https://doi.org/10.1007/978-3-030-17115-5_3

other fields. This statement can be entirely attributed to the study of surface cyclotron waves of various modes.

## 3.1 Dispersion Equation of Surface Cyclotron TM-Modes for Semibounded Plasmas

Here the case of a semi-infinite plane plasma waveguide is considered (see Fig. 3.1). It consists of a plasma with uniform density, which occupies the half-space $z > 0$ and is bordered in the plane $z = 0$ by a dielectric, which has the thickness $a_{\mathrm{d}}$, and the dielectric permittivity $\varepsilon_{\mathrm{d}}$. On the other side, in the space $z \leq -a_{\mathrm{d}}$, the dielectric is bordered by the metal wall of the waveguide, which is assumed to have ideal conductivity. The boundary between the plasma and the dielectric is assumed to be quite sharp, so that the thickness of the transition layer is much less than the penetration depth of surface waves into the plasma. An external constant magnetic field $\vec{B}_0$ is perpendicular to the surface of the plasma and is oriented parallel to the coordinate axis $\vec{z}$. This orientation of the field $\vec{B}_0$, for example, can be realized in the divertor region of magnetic confinement devices for thermonuclear plasma fusion research and as well in plasma electronics devices, when the wall of the plasma filled waveguide structure is made of a ferromagnetic metal.

As it was shown in Chap. 2, in this case the motion of plasma particles can be described by the kinetic Vlasov-Boltzmann equation and their unperturbed distribution function can be taken as Maxwellian. The mirror reflection model is applied for the description of the interaction of plasma particle with the plasma surface, which is often used in solving problems related to diffraction of electromagnetic waves on plasma structures or to interaction of plasma particles with solids [17]. Along the direction of the $\vec{y}$ axis the plasma structure is assumed to be homogeneous. Under

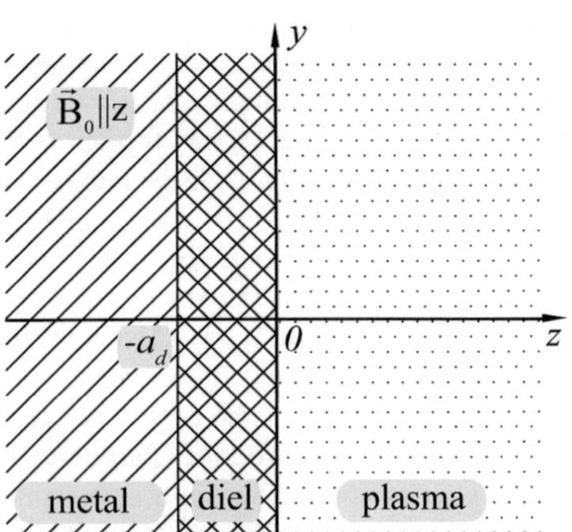


**Fig. 3.1** Schematic of the problem "semibounded plasma-dielectric-metal"

these conditions, one can use the expressions for the dielectric permittivity tensor $\varepsilon_{ik}$ of a gyrotropic plasma that are derived in the previous Chap. 2. It is interesting that it is identical with the expression for the dielectric permittivity tensor $\varepsilon_{ik}$ obtained in the case of an unbounded magnetoactive plasma [1–3].

To describe the electromagnetic field of surface cyclotron TM-modes (SCTM-modes) the set of Maxwell equations is applied. The dependence of the SCTM-mode fields on longitudinal coordinate $x$ and time $t$ is chosen in the form: $E, H \propto \exp(ik_1 x - i\omega t)$, where $k_1$ and $\omega$ are the wave vector component along the $\vec{x}$ axis and the angular frequency of the SCTM-modes, respectively. Here one considers the case of slow waves for which the inequality $\upsilon_{ph}^2|\varepsilon_{ik}| \ll c^2$ is satisfied, where $c$ and $\upsilon_{ph} = \omega/k_1$ are the speed of light and phase velocity of the SCTM-modes, respectively.

The present study is restricted also by the following two conditions. First, it is assumed that the spatial dispersion of the plasma is weak along the normal axis relative to its interface, i.e. along the $\vec{z}$ axis, so that the inequality $k_3 \upsilon_{T\alpha} \ll \omega - s|\omega_\alpha|$ is satisfied, where $k_3$ is the component of the SCTM-mode wave vector along the $\vec{z}$ axis. The second condition is that the space is supposed to be homogeneous along the $\vec{y}$ axis, so that $\partial/\partial y = 0$. It should be pointed out immediately, that the implementation of the first condition leads to the fact that the anti-Hermitian parts of the dielectric permittivity tensor $\varepsilon_{ik}$ of the considered gyrotropic plasma are smaller than the Hermitian parts of this tensor. The second condition allows one to ignore the value of the $E_y$ component of the electric field of these modes as compared with the other components of their electric field.

Under these conditions, the system of Maxwell equations can be separated into two independent subsets. The solution of one of them is of surface type. It describes an $E$-wave (or TM-wave; which means transverse magnetic wave, because the direction of the magnetic field of this wave is perpendicular to the wave vector) with the field components $E_x$, $H_y$, $E_z$ [18]:

$$\begin{aligned} ikH_y &= \mathrm{d}E_x/\mathrm{d}z - ik_1 E_z, \\ \mathrm{d}H_y/\mathrm{d}z &= ikE_x - 4\pi j_x/c, \\ ik_1 H_y &= -ikE_z + 4\pi j_z/c, \end{aligned} \tag{3.1}$$

where $k = \omega/c$ is the so called vacuum wavenumber, because it characterizes electromagnetic waves propagating in vacuum with the angular frequency $\omega$.

From the analysis of this set of equations (3.1) one can see that the functions $E_x$ and $j_x$ are symmetric with respect to the coordinate $z$, and the other three variables ($H_y$, $E_z$ and $j_z$) are asymmetric functions of this coordinate. Therefore, applying a direct Fourier transform to the set of equations (3.1), the fields of the studied SCTM-mode are extended into the area $z < 0$ as follows: the tangential electric field $E_x(z = +0) = E_x(z = -0)$ is even, the normal electric field $E_z$ and the tangential magnetic field $H_y$ are odd. That means, for example, that one can apply the following relation: $H_y(z = +0) = -H_y(z = -0)$ doing integration for the tangential magnetic field component of the SCTM-mode. This allows one to derive the algebraic set of equations for the Fourier coefficients of the SCTM-mode field in the following form:

$$
\begin{aligned}
&ik\varepsilon_{11}E_1 - ik_3H_2 + ik\varepsilon_{13}E_3 = -H_y(0)/(2\pi),\\
&ik\varepsilon_{31}E_1 + ik_1H_2 + ik\varepsilon_{33}E_3 = 0,\\
&ik_3E_1 - ikH_2 - ik_1E_3 = E_x(0)/(2\pi),
\end{aligned} \tag{3.2}
$$

where $E_x(0)$ and $H_y(0)$ are the values of the relevant components of the field of this TM-mode just on the surface of the studied gyrotropic plasma region.

To obtain the dispersion relation for SCTM-modes propagating in such structures, it is necessary to solve (3.2) with respect to the tangential components of the mode's field in the plasma ($z \geq 0$) and dielectric ($-a_\mathrm{d} < z < 0$) regions, respectively. To investigate the field of these modes in the dielectric region one should apply (3.2) by performing the following replacements: $\varepsilon_{11} \to \varepsilon_\mathrm{d}$, $\varepsilon_{33} \to \varepsilon_\mathrm{d}$, $\varepsilon_{13} \to 0$. Then, in the approximation of slow waves, which means validity of the inequality $k_1^2 \gg k^2\varepsilon_\mathrm{d}$, one can easily obtain a homogeneous second order differential equation for the tangential components of the electric and magnetic fields of these TM-modes. This equation can be solved by standard Euler substitution as exponential function $\exp(\pm ik_1z)$. One of the constants of integration obtained for that solution can be determined using the boundary conditions for the tangential component of the SCTM-mode electric field on the interface dielectric-metal. Its value is equal to zero since the metal wall is considered here to be an ideal conductor (as compared with the plasma). Thus, the solution of equations (3.1) in the dielectric medium ($-a_\mathrm{d} < z < 0$) allows one to find the surface impedance $Z_\mathrm{d}(0) = E_x(0)/H_y(0)$ on the interface plasma-dielectric:

$$
Z_\mathrm{d}(0) = -i|k_1|(k\varepsilon_\mathrm{d})^{-1}\tanh(|k_1|a_\mathrm{d}). \tag{3.3}
$$

After setting the tangential component of the electric field of these TM-modes on the interface dielectric-metal (the surface $z = -a_\mathrm{d}$) to zero one can calculate the plasma impedance $Z_\mathrm{pl}(0) = E_x(0)/H_y(0)$ which should be equal to the dielectric impedance on the interface of these media (plane surface $z = 0$). Such procedure corresponds to the application of the following two boundary conditions, which are typical for linear problems in plasma electrodynamics.

- The tangential component of the electric and magnetic fields of a SCTM-mode should be continuous on the interface plasma-dielectric.
- The tangential component of the wave electric field must be equal to zero on the interface dielectric-metal, because the metal is assumed to be an ideal conductor, so the value of its conductivity is considered as infinitely large compared to the plasma conductivity.

Solving the set of equations (3.2) in the plasma region, one can find expressions for the Fourier coefficients of the tangential and normal electric fields of TM-modes, respectively:

$$E_1(k_3) = \frac{ik_1^2 H_y(0)}{\pi \cdot k \cdot \Delta(k_3)}, \quad E_3(k_3) \approx \frac{k_3}{k_1} E_1(k_3). \tag{3.4}$$

Here $\Delta(k_3) = k_1^2 \varepsilon_{11} + 2k_1 k_3 \varepsilon_{13} + k_3^2 \varepsilon_{33}$.

Since the components of the plasma dielectric permittivity tensor $\varepsilon_{ik}$, even under the condition of the absence of collisions between plasma particles, has both Hermitian and anti-Hermitian components one can separate the expression for $\Delta(k_3)$ into real and imaginary parts and write $\Delta(k_3) = \Delta'(k_3) + i\Delta''(k_3)$. Thus, taking into account the small (under condition of weak spatial dispersion $k_3 \upsilon_{T\alpha} \ll \omega - s|\omega_\alpha|$) imaginary part of the denominator $\Delta(k_3)$ in the expression for the Fourier coefficient of the tangential electric field $E_1(k_3)$, one can obtain:

$$1/\Delta(k_3) = (\Delta' - i\Delta'')/[(\Delta')^2 + (\Delta'')^2] \approx 1/\Delta'(k_3) - i\Delta''/|\Delta(k_3)|^2. \tag{3.5}$$

The second term in the expression (3.5) for $1/\Delta(k_3)$ is imaginary. It determines the damping of SCTM-modes. That is why in this section this term will not be taken into account. It will be considered in Sect. 3.2, which is devoted to investigation of mechanisms of mode damping. Thus, further in the present section the imaginary part of $\Delta(k_3)$ is neglected. Since the real part is only considered, the upper index "prime" is omitted to simplify the notation.

Performing the inverse Fourier transform with respect to $E_1(k_3)$ will allow one to calculate the surface impedance of this semibounded plasma $Z_{\text{pl}}(0) = E_x(0)/H_y(0)$. The integral located on the right hand side of the following expression:

$$\int_{-\infty}^{+\infty} E_1(k_3) \exp(ik_3 z) \mathrm{d}k_3 = \frac{ik_1^2 H_y(0)}{\pi \cdot k} \int_{-\infty}^{+\infty} \frac{\exp(ik_3 z) \mathrm{d}k_3}{\Delta(k_3)} \tag{3.6}$$

can be calculated by the theory of residues. Indeed one has to expand it according to the conditions of Jordan's lemma by a curvilinear integral in the upper complex half-plane $\int_{k_3 \to i\infty} \Delta^{-1}(k_3) \exp(ik_3 z) \mathrm{d}k_3$. The expression on the right hand side of equation (3.6) will form an integral over a closed circuit consisting of an infinitely long line, which runs along the axis of real values of the normal component of the wave vector $k_3$ of these TM-modes, and an arc of a semicircle, which is located in the upper complex half-plane of the $k_3$ values, and has infinitely large radius. It should be emphasized that just the circumstance that $k_3$ turns to infinitely large positive imaginary values along this semicircle ($k_3 \to +i\infty$) is the reason that this curvilinear integral is equal to zero. Therefore, on one hand one can add it to the right hand side of equation (3.6) and it does not change the value of this part of equation (3.6). On the other hand this addition allows one to replace the integral over this closed circuit by the sum of residues of the integrand at those values of solutions of the equation $\Delta(k_3) = 0$, which are located inside the specified circuit.

In this case, there is only one such solution for the SCTM-modes, namely:

$$k_3 = i|k_1|\varepsilon_{11}/(\varepsilon_{33} + 2B), \tag{3.7}$$

where $B = \varepsilon_{13}k_1/k_3$ does not depend on the wave vector of the SCTM-mode. Since the normal component of these wave vectors is imaginary, the amplitude of these modes is at the distance $z_* = 1/|k_3|$ only $1/e$-times the value calculated at the plasma-dielectric interface ($z = 0$). That is why it is natural to call this parameter $1/|k_3|$ as penetration depth of the electron SCTM-modes into the plasma.

After calculation of the penetration depth, one can derive the following expression for the plasma impedance on its boundary under the condition of weak spatial dispersion along the $\vec{z}$ axis:

$$Z_{\rm pl}(0) = i|k_1|/[k\sqrt{\varepsilon_{11}(\varepsilon_{33} + 2B)}]. \tag{3.8}$$

By inserting the impedance (3.8) on the right hand side of expression (3.3) one can obtain the dispersion relation for SCTM-modes in this waveguide structure:

$$\sqrt{\varepsilon_{11}(\varepsilon_{33} + 2B)} = \varepsilon_{\rm d}\coth(|k_1|a_{\rm d}). \tag{3.9}$$

The dispersion relation (3.9) is studied here only for the case of electron SCTM-modes. However, the analogous investigation can be done for ion SCTM-modes. Simple analytical expressions for the eigenfrequencies of SCTM-modes at harmonics of the electron cyclotron frequency can be obtained in the limiting cases of wide ($|k_1|a_{\rm d} \gg 1$) and narrow ($|k_1|a_{\rm d} \ll 1$) dielectric layer. Since by definition [1–3], frequencies of cyclotron waves are close to the corresponding cyclotron harmonics, the frequency of the electron SCTM-mode conveniently can be presented in the form $\omega = s|\omega_e|(1 - h_e)^{-1}$, where the parameter $h_e \ll 1$ is small. The physical meaning of $h_e$ is the frequency shift of these modes relative to the corresponding harmonic of the electron cyclotron frequency. To simplify the presentation of further results in this section it is assumed that the plasma is separated from the metal wall by a vacuum layer as dielectric. Therefore, it is characterized by the dielectric constant $\varepsilon_{\rm d} = 1$.

Approximate analytical expressions for the dependence of the frequency of electron SCTM-modes on their wavenumber $k_1$ can be obtained from (3.9), if one uses the advice that is given in monographs [1, 3] for the case of bulk cyclotron waves. Indeed, searching for analytical solutions of the dispersion relation that describes cyclotron waves, it is sufficient to keep the hydrodynamic terms, which are independent on the electron Larmor radius, as well as those terms of the sum over the cyclotron harmonics, which correspond to the harmonic nearby which the solution of the dispersion relation is searched (one calls them "kinetic terms").

Then, in the limiting case of thick dielectric layer ($|k_1|a_{\rm d} \gg 1$) the expressions for the resonant frequency shift for the $s$th harmonic of electron SCTM-modes in the range of long wavelengths ($|k_1|\rho_e \ll 1$) are as follows:

$$h_e \approx -(s^2 - 1)/(2s!)(y_e/2)^{s-1},\ s \geq 2; \tag{3.10}$$

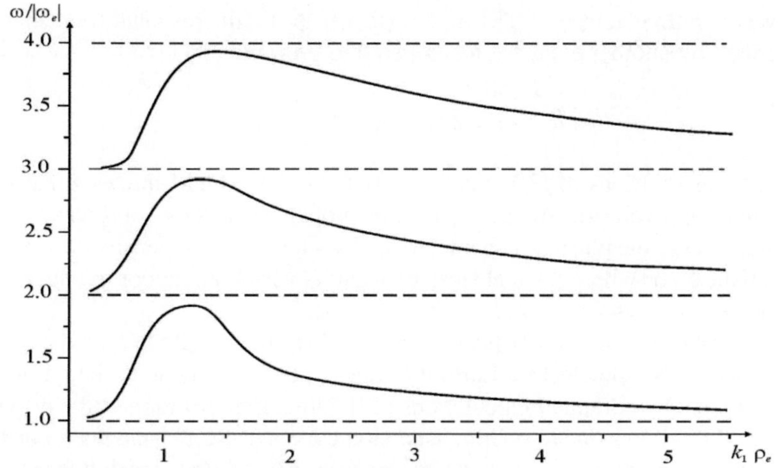


**Fig. 3.2** Dispersion curves of electron SCTM-modes propagating in the first three frequency ranges of electron cyclotron harmonics

where $y_e = k_1^2\rho_e^2/2$. In the range of short wavelengths ($|k_1|\rho_e \gg 1$) the value of the resonant frequency shift $h_e$ is positive:

$$h_e \approx \Omega_e^2\omega_e^{-2}(2\pi y_e^3)^{-1/2}, \quad s \geq 1. \tag{3.11}$$

Thus it is clear that the analytical solutions of the dispersion relation (3.9) in the case of thick dielectric layer (3.10) and (3.11) coincide with the expressions that were obtained in [18] for SCTM-modes which propagate in waveguide structures semibounded plasma-semibounded dielectric.

Results of numerical analysis of the dispersion relation (3.9) are presented in Fig. 3.2 for the following set of waveguide parameters: $a_d/\rho_e = 10, \varepsilon_d = 1, \Omega_e^2/\omega_e^2 = 100$. Analyzing the dispersion curves plotted in Fig. 3.2 one can see that increasing harmonic number $s$ leads to the following changes. First, the point, where the group velocity of these modes is equal to zero, moves to the region of shorter wavelengths and second, the decrease of their frequencies is slower.

Now, in the limiting case of a thin dielectric layer ($|k_1|a_d \ll 1$) [19], the analytical expression for the resonant frequency shift $h_e$ with respect to the electron cyclotron resonance is calculated. For long-wavelength waves ($|k_1|\rho_e \ll 1$) at harmonics of the electron cyclotron frequency one can derive the following expression:

$$h_e = (\Omega_e/\omega_e)^{4/3}(y_e/2)^{(2S-1)/3}(k_d/(s \cdot s!))^{2/3}, \tag{3.12}$$

where $k_d = |k_1|a_d/\varepsilon_d$. Expression (3.12) differs from expression (3.10) first of all by the appearance of the factor $(\Omega_e/\omega_e)^{4/3}a_d^{2/3}$. In the approximation of short-wavelength electron SCTM-modes ($|k_1|\rho_e \gg 1$), the resonant frequency shift is described by another expression compared to Formula (3.11):

$$h_e = (\Omega_e/\omega_e)^{4/3}\left(k_{\rm d}^2/\pi\, s^2 y_e^2\right)^{1/3}. \qquad (3.13)$$

Analytical solutions of (3.12) and (3.13) show that SCTM-modes at harmonics of the electron cyclotron frequency cannot propagate in those plasma waveguide structures, where the plasma is characterized by a low value of density (where $\Omega_e^2 \ll \omega_e^2$ is satisfied), or where the dielectric constant is very high and/or the thickness of the dielectric layer is very low, i.e. when $k_{\rm d} \to 0$.

For an arbitrary relation between the wavelength $\lambda = 2\pi k_1^{-1}$ of the electron SCTM-modes and the electron Larmor radius $\rho_e$, their dispersion relation only can be investigated by computer calculations [19]. Numerical solution of the dispersion equation (3.9) allows one to obtain dispersion curves of SCTM-modes for arbitrary values of the parameters of the plasma waveguide. At first, the influence of the dielectric layer thickness on dispersion of the electron SCTM-modes is studied. The curves obtained for the case of electron SCTM-modes at the first three harmonics of the electron cyclotron frequency are plotted in Fig. 3.3. It shows the frequency dependence of the electron SCTM-modes, normalized to the absolute value of the electron cyclotron frequency, on the wavenumber, which is normalized to the electron Larmor radius. The case of very dense plasma ($\Omega_e^2 = 10^3\omega_e^2$) and vacuum layer ($\varepsilon_{\rm d} = 1$) is under the consideration. The curves marked with asterisks correspond to the cases of different values of the thickness of the dielectric layer, namely $10^{-3}\rho_e$ (solid line) and $10^{-6}\rho_e$ (dashed line). The horizontal axis has a logarithmic scale. Numerical study of equation (3.9) proves that if the thickness of the dielectric layer decreases, the dispersion curve of SCTM-modes propagating in a plasma with weak spatial dispersion ($k_1 a_{\rm d} \gg 1$), becomes approximately a straight horizontal line. With increasing number of cyclotron harmonics the maxima of the dispersion curves of SCTM-modes are shifted to the shorter wavelength range and the frequency of electron SCTM-modes comes closer and closer to the electron cyclotron resonance at the corresponding harmonic $\omega = (s-1)|\omega_e|$, $s \geq 2$.

In the following the influence of the value of the dielectric constant $\varepsilon_{\rm d}$ on the dispersion of electron SCTM-modes is investigated. The numerical calculations show that the dispersion of these modes is approximately a straight horizontal line in the range of long wavelengths for waveguide structures with the parameter $\Omega_e^2 = 10^3 \cdot \omega_e^2$, if $k_{\rm d} \leq 10^{-5}$. Both reducing of the thickness of the dielectric layer $a_{\rm d}$ and increasing of the value of its dielectric permittivity $\varepsilon_{\rm d}$ equally lead to a decrease of the frequency of SCTM-modes and the dispersion curves asymptotically approach to the corresponding resonance lines $\omega = (s-1)|\omega_e|$.

The effect of the magnitude of the dielectric constant of the dielectric layer that separates the plasma from the metal walls of the waveguide on the dispersion of electron SCTM-modes at the second harmonic is illustrated in Fig. 3.4. It presents an analysis of the case of a dense plasma ($\Omega_e^2 = 10^3\omega_e^2$) and the value of the thickness of the dielectric layer was assumed to be $10^{-4}\rho_e$. The dispersion curves plotted in

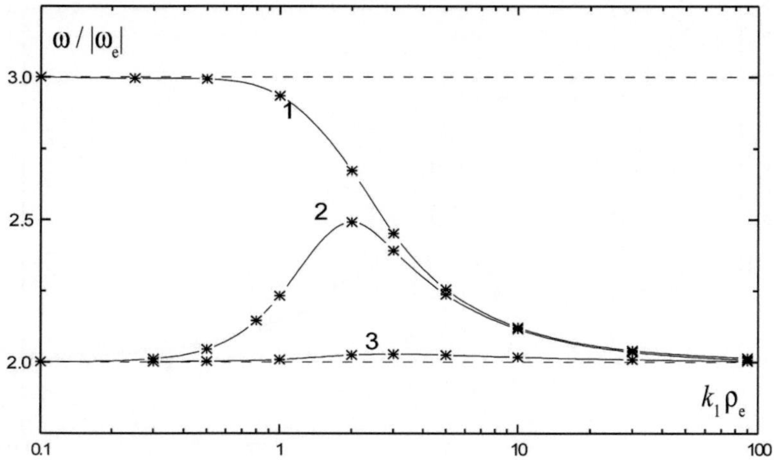


**Fig. 3.3** Dependence of the frequency of electron SCTM-modes on the wavenumber for different thickness of the dielectric coating of the waveguide metal wall

Fig. 3.4 for the cases of dielectric coatings with $\varepsilon_d$ =1, $\varepsilon_d$ =10 and $\varepsilon_d$ =100 are marked with the numbers 1, 2 and 3, respectively. It can be seen that the amplitude of the electron SCTM-modes, as well as their frequency decreases with increasing value of $\varepsilon_d$. In general, an increase of the parameter $k_{d,}$ which characterizes the dielectric properties of the studied waveguide structure, means that the problem turns into the problem of SCTM-modes propagating in the structure of the form semibounded plasma-semibounded dielectric [18]. Thus, it was found that the range of existence of electron SCTM-modes in structures of the form plasma-dielectric-metal is reduced if $k_d$ is decreasing.

If one compares the dispersion curves, which are shown in Figs. 3.2, 3.3 and 3.4 for the case of a semibounded plasma-semibounded dielectric interface, with dispersion curves that are plotted in [1–3] for longitudinal potential bulk cyclotron waves, which are described by the dispersion relation $\varepsilon_{11}(\omega, \vec{k}) = 0$, one can see that they are similar. Analysis of the factors $\varepsilon_{11}(\omega, k_1)$ and $\varepsilon_{33} + 2B$ that determine the dispersion relation of SCTM-modes (3.9), shows that they have the following magnitudes: $\varepsilon_{11} \ll 1$ and $\varepsilon_{33} + 2B \gg 1$. Satisfaction of these inequalities explains the mentioned similarity of the dispersion curves for potential bulk electron cyclotron waves and surface electron cyclotron waves with TM-polarization in the case of the waveguide structure discussed here.

In addition to the analytical comparison of electron SCTM-modes with potential bulk cyclotron waves, the evaluation of absolute values of the factors indicated above, namely the first diagonal component of the dielectric permittivity tensor $\varepsilon_{11}$ and the sum $\varepsilon_{33} + 2B$, which describe the electromagnetic properties of the plasma, allows one to determine from expression (3.7) the relationship between the absolute values of components of the SCTM-mode wave vector $k_1$ and $k_3$:

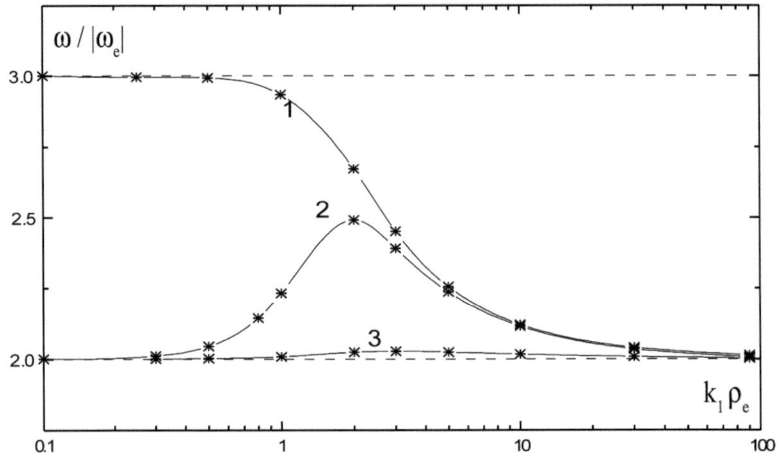


**Fig. 3.4** Dependence of the frequency of electron SCTM-modes on the wavenumber for different values of the dielectric constant of the dielectric waveguide coating

$$|k_3/k_1| = \left| \frac{\varepsilon_{11}}{\varepsilon_{33} + 2B} \right| \ll 1. \tag{3.14}$$

Analysis of the inequality (3.14) proves that the penetration depth $1/|k_3| = 1/k_\perp$ of SCTM-mode fields into the plasma region is larger than the wavelength of these modes.

Numerical analysis of the expression (3.14) confirms the obtained analytical results, that the magnitude $|k_3/k_1|$ is really much less than unit. In Fig. 3.5, one can see that the local maximum of this ratio is located nearby the inflection point of the dispersion curve $\omega = \omega(k_1)$ and that the value of this local maximum decreases with increasing plasma density.

The case of ion SCTM-modes from a mathematical point of view practically does not differ from the case of electron SCTM-modes. This is due to the fact that the hydrodynamic terms in the dielectric tensor components are symmetrical both in the case of electron cyclotron and ion cyclotron TM waves. Therefore, the approximate analytical formulas (3.10)–(3.13) can also be used in first approximation to the case of ion SCTM-modes, simply by changing the "electron" indices by "ion" indices.

## 3.2 Damping of Surface Electron Cyclotron TM-Modes

Damping of electron SCTM-modes in the studied waveguide structures can happen due to the following reasons. The first one is the presence of collisions between the particles of the plasma preventing a synchronized collective motion of plasma particles in an external static magnetic field. The number of collisions per unit of

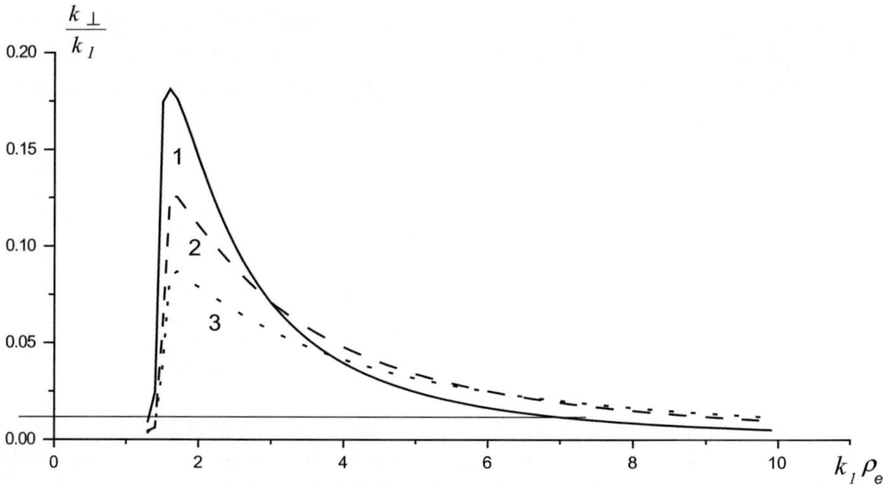


**Fig. 3.5** Ratio of SCTM-mode wavelength normalized to $2\pi k_\perp^{-1}$ versus the product $k_1\rho_e$. Solid, dashed and dotted lines show results calculated for the SCTM-mode at the second electron cyclotron harmonic in the cases $\Omega_e^2/\omega_e^2 = 10$, 50 and 100, respectively

time is characterized by the effective collision frequency $\nu_\alpha$. The second reason is connected with the interaction of plasma particles with the plasma interface. Since the amplitude of surface waves is of maximum value nearby the plasma boundary, collisions of plasma particles with the plasma boundary influence the collective motion of these particles as well.

To determine the damping rate $\delta_{\rm col}$ caused by collisions between plasma particles one can take into account the value $\nu_\alpha$ in the expressions for the components of the plasma conductivity tensor $\sigma_{ik}$. Doing this it is necessary to consider the fact that small corrections to the imaginary frequency of SCTM-modes also lead to the appearance of an imaginary part in the magnitude of the wavenumber $k_1$. This means that from a mathematical point of view, it is necessary to make the following two replacements in the dispersion relation (3.9): $k_1 \to k_1 + i\delta_{\rm col}$ and $\omega \to \omega + i\nu_\alpha$.

This specified change of the wave frequency is the result of searching the solution of the kinetic Vlasov-Boltzmann equation. This changing can be done only for the frequency, which is present in the conductivity tensor $\sigma_{ik}$ but not for the frequency, which has to be used for determination of the plasma dielectric permittivity tensor $\varepsilon_{ik} = \delta_{ik} + 4\pi\, i\, \sigma_{ik}/\omega$. Here $\delta_{ik}$ is the Kronecker symbol. As a last remark about this problem one should mention, that a small correction $i\nu_\alpha$ exerts the greatest impact on the frequency shift $h_\alpha$ of SCTM-modes with respect to the corresponding harmonic cyclotron frequency, because by the definition of cyclotron waves these are waves, whose frequency is close to harmonics of the cyclotron frequency. In the following, the indicated substitution $h_\alpha \to (\omega + i\nu_\alpha - s|\omega_\alpha|)/\omega = h_\alpha[1 + i\nu_\alpha/(\omega h_\alpha)]$ of the electron SCTM-mode frequency shift is introduced and most attention is payed just to these terms in the dispersion relation.

Thus, taking the imaginary part of the dispersion equation (3.9), which is caused by the presence of collisions between plasma particles, it is quite simple to calculate the following expression for the collisional damping rate of electron SCTM-modes:

$$\frac{\delta_{\text{col}}}{|k_1|} = \xi \frac{\nu_e}{\omega h_e}, \tag{3.15}$$

where

$$\xi = \begin{cases} (2s-2)^{-1}, & h_e < 0,\ y_\alpha \ll 1, \quad s \geq 2, \\ -3/(4s-2), & h_e > 0,\ y_e < 1 < y_i < 1,\ s \geq 1, \\ -1/3, & h_e > 0,\ y_\alpha \gg 1, \quad s \geq 1. \end{cases}$$

One can see from analysis of expression (3.15) that if the frequency shift $h_e$ turns to zero (in other words, if the wavelength of electron SCTM-modes becomes extremely short or long), the value of their collisional damping rate strongly increases.

Another reason for the damping of SCTM-modes in the considered waveguide structure is the spatial dispersion of the plasma, that is caused by the thermal motion of the plasma particles, which leads to collisions of plasma particles not only among themselves but also with the plasma interface. Calculating the relevant anti-Hermitian parts in the expressions of dielectric permittivity tensor components calculated in the previous Chap. 2 for a semibounded gyrotropic plasma, one can estimate the corresponding damping rates of electron SCTM-modes. Their values are relatively small, because the inhomogeneity of the SCTM-mode fields along the direction $\vec{B}_0 \| \vec{z}$ (which is normal to the plasma interface) is weak. Analytical expressions for these rates have been obtained in [18] in the case of the mirror reflection model of plasma particles from the interface plasma-dielectric.

It is clear that the accuracy of the description of the interaction of the plasma particles with the plasma boundary can be determined and checked by experimental verification of the obtained results. However, the absence of such experiments leads one to the necessity to solve this problem theoretically, which means that one can only compare the analytical results obtained for different models of this interaction. The results of solving many problems of electrodynamics of bounded plasmas [17, 18] only weakly depend on the choice of the model describing the interaction between plasma particles and the plasma boundary. That is why for the theoretical description one has to take into account difficulties during searching for solutions of the Vlasov-Boltzmann kinetic equation and after that make the choice of the convenient model of the plasma particles-plasma surface interaction. In the case when an external static magnetic field is oriented perpendicularly to the plasma-dielectric interface, the choice of mirror reflection of plasma particles from the interface greatly simplifies the problem of finding a solution of the kinetic equation, i.e. the kinetic equation has the same solution as in the case of an unbounded plasma. Thus, under these conditions the application of the mirror reflection model leads to a simple, linear relation between the Fourier coefficients of the electric current density and the components of the electric field of various electromagnetic waves:

$$j_i(\vec{k}) = \sigma_{ij}(\vec{k})E_j(\vec{k}), \tag{3.16}$$

where $\sigma_{ij}(\vec{k})$ is the plasma electrical conductivity tensor, which has the same form as in the case of unbounded gyrotropic plasmas (see Chap. 2). Thus, even for the case when collisions between plasma particles are absent (collisional frequency $\nu_\alpha = 0$) the conductivity tensor $\sigma_{ik}$ nevertheless has anti-Hermitian components. Therefore, as it was shown in Sect. 3.1, to carry out the calculation of the impedance of a semibounded plasma on its interface in order to derive the damping rate of electron SCTM-modes caused by the interaction between plasma particles and the interface, one can take into account the imaginary part of the right hand side of $\Delta(k_3)$ in (3.6). Then (3.6) can be rewritten as follows:

$$\frac{E_x(0)}{H_y(0)} = \frac{ik_1^2}{\pi \cdot k} \int\limits_{-\infty}^{+\infty} \frac{[\Delta'(k_3) - i\Delta''(k_3)]\mathrm{d}k_3}{|\Delta(k_3)|^2 \cdot \exp(-ik_3 z)}|_{z=0} \approx \frac{i|k_1|}{k\sqrt{\varepsilon_{11}(\varepsilon_{33} + 2B)}} + I_{\mathrm{kin}}, \tag{3.17}$$

where $I_{\mathrm{kin}} = \frac{k_1^2}{\pi \cdot k} \int_{-\infty}^{+\infty} \frac{\Delta''(k_3)\cdot \mathrm{d}k_3}{|\Delta(k_3)|^2 \cdot \exp(-ik_3 z)}|_{z=0}$. The term $I_{\mathrm{kin}}$ determines the value of this kinetic damping rate, which is mathematically connected with the integration of the kinetic equation. This damping can be considered as analogous to Landau damping, however, for such surface waves, their properties require to solve the kinetic equation. This statement is based on the fact that the appearance of this term is also caused by the presence of anti-Hermitian components of the dielectric permittivity tensor obtained for the case of a gyrotropic plasma and by the necessity to calculate an improper integral of the integrant function that has a singularity in its denominator. Taking into account the parity of the integrand in the expression for $I_{\mathrm{kin}}$, it can be rewritten in the following integral form:

$$I_{\mathrm{kin}} = \frac{2k_1^2}{\pi \cdot k} \int\limits_0^{+\infty} \frac{\Delta''(k_3)}{|\Delta(k_3)|^2} \mathrm{d}k_3. \tag{3.18}$$

Calculation of the integral (3.18) can be carried out separately, using the approximations for the cases of weak and strong spatial dispersion of the gyrotropic plasma along the $\vec{x}$ axis. The term "weak spatial dispersion" means that the wavelength in the plasma is much larger than the characteristic spatial scale of plasma particle motion. If the plasma is magnetized, this scale is the Larmor radius $\rho_\alpha$ of gyration of the charged plasma particles around the force lines of the external static magnetic field. Thus in this limiting case, the argument of the modified Bessel functions [20] included in the analytical expression of the plasma dielectric permittivity tensor $\varepsilon_{ij}$ is small: $y_\alpha^2 = k_1^2\rho_\alpha^2/2 \ll 1$. In the case of strong spatial dispersion of the magnetoactive plasma the wavelength is small compared to the Larmor radius: $y_\alpha^2 \gg 1$.

The method of calculating integrals, which are of type (3.18), is proposed and justified in [21]. Since anti-Hermitian components of the dielectric permittivity tensor

$\varepsilon_{ik}$ are proportional to the error integral $W(x_s)$ (for details see Sect. 2.2, here $x_s = (\omega - s\omega_\alpha)/(k_3 \upsilon_{T\alpha})$), for the integration of $I_{\text{kin}}$ one can firstly change integration over the variable $dk_3$ by the new variable $dx_s$. Secondly, the original integral (3.18) should be represented as the sum of two integrals $I_{\text{kin}} = I_{\text{kin1}} + I_{\text{kin2}}$. Doing this one can break the limits of integration, so the integral $I_{\text{kin1}}$ can be integrated in the range from zero to unit, and the integral $I_{\text{kin2}}$ in the range from unit to infinity. This will allow one to apply asymptotes of the integrand functions in the limits $x_s \ll 1$ and $x_s \gg 1$ for the approximate calculation of the integrals $I_{\text{kin1}}$ and $I_{\text{kin2}}$, respectively.

Thus, in the case of weak spatial dispersion of the plasma ($y_\alpha = k_1^2 \rho_\alpha^2/2 \ll 1$) along the propagation direction of SCTM-modes one can derive the following approximate expressions for the indicated imaginary parts of the dispersion relation of these modes:

$$I_{\text{kin1}} \approx \frac{k_1 \upsilon_{T\alpha} \omega}{\Omega_\alpha^2}, \quad I_{\text{kin2}} \approx \frac{2\omega^3 |h_\alpha|}{\sqrt{\pi}\,\Omega_\alpha^2 k_1 \upsilon_{T\alpha}} \left| \ln\left( \frac{|h_\alpha|}{\sqrt{y_\alpha}} \right) \right|. \tag{3.19}$$

If one separates in the wavenumber $k_1$ a small imaginary term $k_1 \to k_1 + i\delta_{\text{kin}}$, here $|\delta_{\text{kin}}| \ll |k_1|$, one can find the damping rates of electron SCTM-modes, which are caused by collisions between plasma particles and plasma boundary (i.e. kinetic damping rate) in the limiting case of weak spatial dispersion along the interface of a gyrotropic plasma ($y_\alpha \ll 1$):

$$\frac{\delta_{\text{kin}}}{k_1} \approx -\frac{\omega_e^6}{\Omega_e^6}. \tag{3.20}$$

In the case of strong spatial dispersion ($y_\alpha = k_1^2 \rho_\alpha^2/2 \gg 1$) along the $\vec{x}$ axis (direction of propagation of the investigated modes), the approximate value of $I_{\text{kin1}}$ remains the same as it was in the case of weak spatial dispersion of the gyrotropic plasma [see expression (3.19)], but the expression $I_{\text{kin2}}$ changes to:

$$I_{\text{kin2}} \approx \frac{2\omega h_\alpha}{\sqrt{\pi} k_1 \upsilon_{T\alpha}} \left| \ln\left( \frac{h_\alpha}{\sqrt{y_\alpha}} \right) \right|. \tag{3.21}$$

Therefore, it leads to some modification of the general expression of the kinetic damping rate as compared to the previous case of weak spatial dispersion:

$$\frac{\delta_{\text{kin}}}{k_1} \approx -\frac{s^2 h_\alpha \omega_\alpha^2}{3\Omega_\alpha^2 \sqrt{y_\alpha}}. \tag{3.22}$$

Comparing the magnitudes of the damping rates $\delta_{\text{col}}$ caused by collisions between plasma particles and the damping rates $\delta_{\text{kin}}$ determined by the interaction of plasma particles with the plasma boundary (kinetic damping), one can conclude that $\delta_{\text{col}} > \delta_{\text{kin}}$ in both limiting cases. The reason for this is that the collisional damping $\delta_{\text{col}} \propto 1/h_\alpha$ and at the same time the kinetic damping is not strong because of weak inhomogeneity of the electron SCTM-mode field along the coordinate normal to the plasma interface.

It should be noted also that the sign of the damping rates of electron SCTM-modes is negative in all considered limiting cases. This is a consequence of the fact that the phase and group velocities of these modes in the ranges of long ($y_\alpha \ll 1$) and short ($y_\alpha \gg 1$) wavelengths are directed in mutually opposite directions (in other words the dispersion of electron SCTM-modes is reverse) in the case of a dense plasma ($\Omega_e^2 \gg \omega_e^2$). In addition, one can understand that any damping of electromagnetic waves takes place along the direction of the group velocity, i.e. the direction of transferring the wave power.

Therefore, in gyrotropic plasmas with low level of collisions, electron SCTM-modes are weakly damped and the magnitude of their damping rate is mainly determined by collisions between plasma particles. The results presented above are obtained for the case of propagation of SCTM-modes along the interface of semi-bounded plasmas, however, they are approximately the same as in the case of a plasma layer with a thickness of finite size.

## 3.3 Effect of Non-uniform Plasma Particle Density and Finite Transverse Plasma Size on Spectrum of Surface Electron Cyclotron TM-Modes

Here the influence of finite transverse size of the plasma filling of a planar metal waveguide on the dispersion properties of electron SCTM-modes is considered. The waveguide walls have a protective dielectric coating. These theoretical studies will be carried out for a wide range of parameters of the dielectric medium, which coats the waveguide wall, and magnitude of the external static magnetic field oriented perpendicularly to the plasma-dielectric interface. To do this, the following geometry of the planar waveguide filled with plasma is considered (see Fig. 3.6). The metal walls of the waveguide are characterized by perfect electric conductivity, they are located in the areas $z < -a_\mathrm{d}$ and $z > a$, where $a = a_\mathrm{pl} + a_\mathrm{d}$. Both of them are covered on the inner side with a dielectric layer with thickness $a_\mathrm{d}$ and dielectric permittivity $\varepsilon_\mathrm{d}$. The gap between these dielectric coatings is filled with the uniform plasma slab with thickness $a_\mathrm{pl}$. The plasma consists of electrons and one kind of ions. This plasma filled waveguide is assumed to be uniform along the direction of the $\vec{y}$ axis. Electron SCTM-modes propagate along the $\vec{x}$ axis.

To find the spatial distribution of the SCTM-mode fields and the dispersion of these modes one uses the same approaches, which were proposed in the previous section. In order to solve the Maxwell equations, the Fourier method is used. Then the set of differential equations is reduced to a system of algebraic equations for the Fourier coefficients of the SCTM-mode fields in the plasma region:

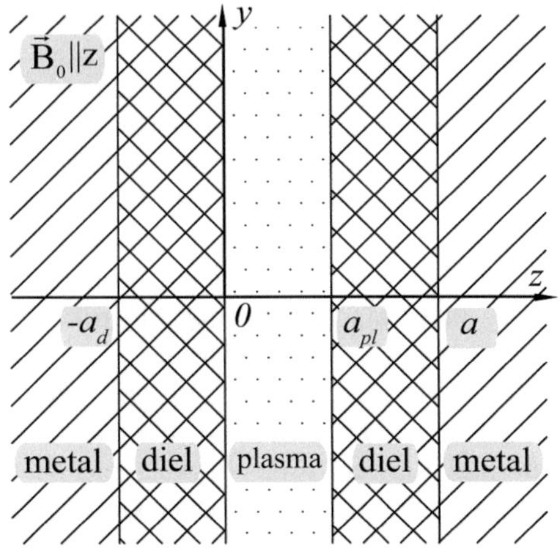


**Fig. 3.6** Schematic of the problem of studying the effect of finite transverse plasma size on SCTM-mode spectrum

$$
\begin{aligned}
&(ik\varepsilon_{11}E_1 - i\alpha_n H_2 + ik\varepsilon_{13}E_3)2a_{\mathrm{pl}} = (-1)^n H_y(a_{\mathrm{pl}}) - H_y(0),\\
&k\varepsilon_{31}E_1 + k_1 H_2 + k\varepsilon_{33}E_3 = 0,\\
&i\alpha_n E_1 - ikH_2 - ik_1E_3 = -\frac{(-1)^n E_x(a_{\mathrm{pl}}) - E_x(0)}{2a_{\mathrm{pl}}},
\end{aligned}
\tag{3.23}
$$

where $\alpha_n = \pi n/a_{\mathrm{pl}}$ are Fourier variables in the sum of Fourier harmonics over the indices $n = 0, \pm1, \pm2, \ldots$. The expressions for the components of the dielectric tensor are similar to those used in the case of semibounded plasmas obtained in Sect. 2.1 with the only difference, that the axial component of the wave vector $k_z$ should be replaced by the parameter $\alpha_n = \pi n/a_{\mathrm{pl}}$. Such replacing is usual for any plasma waveguide with finite value of its transverse size [21].

The set of equations (3.23) can also be used to find the distribution of the SCTM-mode fields in the regions of dielectric layers, by doing as previously the following replacements: $\varepsilon_{11} \to \varepsilon_{\mathrm{d}}$, $\varepsilon_{33} \to \varepsilon_{\mathrm{d}}$, $\varepsilon_{13} = \varepsilon_{31} \to 0$ and $a_{\mathrm{pl}} \to a_{\mathrm{d}}$.

Using the inverse Fourier integral transformation, one obtains the dependence of the components of these mode fields along the transverse coordinate of the studied waveguide structure. It allows one to calculate the tangential components of the electric and magnetic fields of SCTM-modes on the boundaries between the plasma filling and these two dielectric layers. After that one can obtaine expressions for the impedance on the corresponding interface of the studied plasma filled waveguide:

$$
\begin{aligned}
Z_{\mathrm{d}}(-a_{\mathrm{d}}) &= iA_2\left\{\frac{Z_{\mathrm{d}}(0)\cosh(|k_1|a_{\mathrm{d}}) + A_2\sinh(|k_1|a_{\mathrm{d}})}{Z_{\mathrm{d}}(0)\sinh(|k_1|a_{\mathrm{d}}) + A_2\cosh(|k_1|a_{\mathrm{d}})}\right\},\\
Z_{\mathrm{pl}}(0) &= iA_1\left\{\frac{Z_{\mathrm{pl}}(a_{\mathrm{pl}})\cosh(|k_1|ba_{\mathrm{pl}}) + A_1\sinh(|k_1|ba_{\mathrm{pl}})}{Z_{\mathrm{pl}}(a_{\mathrm{pl}})\sinh(|k_1|ba_{\mathrm{pl}}) + A_1\cosh(|k_1|ba_{\mathrm{pl}})}\right\},\\
Z_{\mathrm{d}}(a_{\mathrm{pl}}) &= iA_2\left\{\frac{Z_{\mathrm{d}}(a)\cosh(|k_1|a_{\mathrm{d}}) + A_2\sinh(|k_1|a_{\mathrm{d}})}{Z_{\mathrm{d}}(a)\sinh(|k_1|a_{\mathrm{d}}) + A_2\cosh(|k_1|a_{\mathrm{d}})}\right\},
\end{aligned}
\tag{3.24}
$$

where

$$A_1 = \frac{|k_1|}{k\sqrt{\varepsilon_{11}(\varepsilon_{33}+2B)}}, \quad A_2 = \frac{|k_1|}{k\varepsilon_d}, \quad b = \sqrt{\frac{\varepsilon_{11}}{\varepsilon_{33}+2B}}.$$

To derive the dispersion relation for SCTM-modes propagating in such waveguide it is appropriate to equate the impedances calculated on the two boundaries of the plasma filling: $Z_{pl}(0) = Z_d(0)$ and $Z_{pl}(a_{pl}) = Z_d(a_{pl})$. Since it is assumed that the metal walls have an ideal electric conductivity, the dielectric surface impedance on the boundaries of the waveguide metal walls must be zero: $Z_d(-a_d) = 0$ and $Z_d(a) = 0$. Using these boundary conditions, one gains the dispersion relation for the studied waveguide structure. It reads as follows:

$$\sqrt{\varepsilon_{11}(\varepsilon_{33}+2B)} \cdot \tanh\big(|k_1|ba_{pl}\big) = \varepsilon_d \coth(|k_1|a_d). \tag{3.25}$$

Considering this (3.25) one can conclude that the propagation of SCTM-modes strongly depends on the values of transverse sizes of the plasma filling layer.

In the following, this equation is analyzed in details. First of all, since the ion Larmor radius is much larger than the electron Larmor radius, one should recognize that the limited transverse size of the plasma layer mostly influences the electromagnetic waves propagating at the ion cyclotron frequency. Thus the effect of the plasma layer thickness on the dispersion properties of electron SCTM-modes can be essential for solid state plasmas and/or meta-plasma structures that can be utilized in compact devices of microelectronics. Second, analytical expressions for the solutions of equation (3.25) can be conveniently expressed in the form: $\omega = s\omega_e/(1-h_e)$, where $|h_e| \ll 1$. They can be obtained analytically only in the limiting cases of long ($|k_1|\rho_e \ll 1$) and short ($|k_1|\rho_e \gg 1$) wavelengths using the approximations of thick $\big(|k_1|ba_{pl} \gg 1\big)$ or thin $\big(|k_1|ba_{pl} \ll 1\big)$ plasma layer and thick ($|k_1|a_d \gg 1$) or thin ($|k_1|a_d \ll 1$) dielectric layers.

In the limiting case of thick plasma and dielectric layers, the dispersion relation (3.25) reduces to the structure of the dispersion relation, which was derived for the case of a semibounded plasma—semibounded dielectric interface. This plasma filled waveguide structure was discussed in the previous Sect. 3.1, so that one does not need to repeat these results in the present section in detail. This case is properly determined by formulas (3.5) and (3.6).

In the limiting case of a thin plasma layer and thick dielectric layers the frequency of electron SCTM-modes is determined through the following expressions for the resonant frequency shift $h_e$:

$$h_e = \left(\frac{\Omega_e}{\omega_e}\right)^2 \frac{s^2 y_e^{s-1}}{2^{s+1}s!}\frac{k_{pl}}{\varepsilon_d}, \quad y_e \ll 1,$$
$$h_e = \left(\frac{\Omega_e}{\omega_e}\right)^2 \frac{s^2}{\sqrt{\pi}(2y_e)^{3/2}}\frac{k_{pl}}{\varepsilon_d}, \quad y_e \gg 1, \tag{3.26}$$

where $k_{pl} = k_1 a_{pl}$.

In the limiting case of a thin plasma layer and thin dielectric layers the analytical expressions for the resonance frequency shift of electron SCTM-modes can be written in the following form:

$$h_e = \left(\frac{\Omega_e}{\omega_e}\right)^2 \frac{s^2 y_e^{s-1}}{2^{s+1} s!} k_{\text{pl}} k_{\text{d}}, \; y_e \ll 1,$$
$$h_e = \left(\frac{\Omega_e}{\omega_e}\right)^2 \frac{s^2 k_{\text{pl}} k_{\text{d}}}{\sqrt{\pi}(2y_e)^{3/2}}, \; y_e \gg 1, \tag{3.27}$$

where $k_{\text{d}} = k_1 a_{\text{d}}/\varepsilon_{\text{d}}$.

Analysis of the asymptotic solutions (3.26) and (3.27), obtained for the dispersion relation (3.25), shows that a decrease of the thickness of both dielectric and plasma layers and also an increasing value of the dielectric permittivity of the dielectric coating separating the plasma volume from the metal waveguide walls leads to deterioration of the conditions of electron SCTM-mode propagation. This means that decreasing the waveguide dimension $a$ leads to an increase of the damping rates of these modes caused by collisions between the plasma particles, because the smaller the value of the resonance shift is, the larger is the value of the damping rates (see Sect. 3.2).

As it is known from the general theory of surface waves [18, 21], in planar plasma layers generally speaking, two types of eigenwaves can propagate. First, a symmetric mode for which the tangential components of the electromagnetic field at the boundaries of the plasma layer behave as $E_x(0) = E_x(a_{\text{pl}})$ and $H_y(0) = -H_y(a_{\text{pl}})$ can exist. Second, an anti-symmetric mode with the opposite behavior of its fields, namely $E_x(0) = -E_x(a_{\text{pl}})$ and $H_y(0) = H_y(a_{\text{pl}})$ is present. Their dispersion properties are different, and their dispersion relations have different forms. The symmetric mode dispersion is described by (3.25). Unlike that, the anti-symmetric mode dispersion relation is characterized by the following replacement: on the left-hand side the hyperbolic tangent $\tanh\big(|k_1| b a_{\text{pl}}\big)$ as a factor, which characterizes the thickness of the plasma filling, must be replaced by another function, namely, the hyperbolic cotangent $\coth\big(|k_1| b a_{\text{pl}}\big)$. Then for symmetric waves propagating in a thin plasma filling layer $(a_{\text{pl}} k_1)^2 \ll |(\varepsilon_{33} + 2B)/\varepsilon_{11}|$ the form of the dispersion relation (3.25) can be simplified for analytical studies in the following way:

$$\varepsilon_{11} \cdot a_{\text{pl}} k_1 \approx -\varepsilon_{\text{d}} \cdot \coth(|k_1|\, a_{\text{d}}). \tag{3.28}$$

Thus for the case of anti-symmetric mode propagation in thin layers of plasma filling the dispersion relation of the TM-modes has the form $\varepsilon_{11} = 0$, which coincides with the equation for longitudinal bulk potential cyclotron waves. That means that waves with TM polarization cannot propagate in the form of anti-symmetric modes in such waveguides. Therefore, the surface waves in the case of a plasma filling that is described by the model of a planar layer always have the form of symmetrical modes only.

For arbitrary value of thickness of both dielectric and plasma layers the dispersion relation (3.25) can be solved only by numerical methods. It has been proved numerically that decreasing plasma filling thickness leads to decreasing electron SCTM-mode frequencies and in this case the dispersion curves approach to the values of the corresponding resonant electron cyclotron frequency harmonics. The dependence of the dispersion of SCTM-modes on the thickness of the plasma layer considered here is similar to the dependence of the dispersion of these modes on the thickness of the dielectric layer in a waveguide structure that is modelled by a semibounded plasma-dielectric layer-metal wall structure (see Sect. 3.1).

The dependence of the electron SCTM-mode frequencies on the magnitude of the external static magnetic field $\vec{B}_0$ is more complicated. Variation of the $\vec{B}_0$ value primarily affects the frequency spectrum of the modes that belong to the long-wavelength regime. Namely, increasing value of the external magnetic field leads to decreasing frequency of electron SCTM-modes and therefore the region of the wavenumbers of these modes, where damping does not affect essentially the SCTM-mode propagation, becomes narrower. In the region of long wavelengths, wavenumber ranges exist where the reverse dispersion of electron SCTM-modes changes to normal dispersion.

Moreover, in waveguides of this geometry with thin dielectric coating of the inner surface of their metal walls, the rate of energy transfer by the SCTM-modes significantly increases along the waveguide structure. This is explained by the increasing group velocity of these modes in the middle part of the range of their wavelengths as compared with the case of the model semibounded plasma-dielectric coating. In this case, the value of the velocity of energy transfer increases as well with increasing plasma layer thickness and the number of electron cyclotron harmonics. Thus, changing the thickness of the plasma layer can be utilized for effective regulation of the power transfer rate of electron SCTM-modes along the axis of the waveguide.

Numerical analysis of the dispersion equation (3.25) allows one to study the influence of the magnitude of the applied external static magnetic field on the frequency of the electron SCTM-modes. Results of this analysis are presented in Fig. 3.7. This analysis has been done for the first five electron cyclotron harmonics using the dimensionless parameter $Z = \Omega_e^2 \omega_e^{-2}$. To demonstrate the influence of this parameter on the form of dispersion curves three $Z$ values have been chosen.

After analysis of the curves presented in Fig. 3.7, it is clear that a decrease of the plasma particle density leads to the fact that the frequency of electron SCTM-modes decreases. This means that the dispersion curves approach to the frequency of the corresponding electron cyclotron resonance harmonic in both regions of long and short wavelengths of these modes. The regions, where SCTM-mode propagation is possible without essential damping, become narrower and narrower. Doing that, the conditions for the existence of these electron TM-modes will be violated, because they can be realized in the case of high density of the plasma filling and a large penetration depth of the electromagnetic field of these modes into the plasma. These conditions are necessary to preserve the basic starting assumption of a weak spatial dispersion of the plasma along the direction of the external static magnetic field. It should be remembered, that exactly the same approach has been applied in the previous Chap. 2 for derivation of the expressions for the components of the plasma dielectric permittivity tensor that is used in this chapter. In the dispersion curves

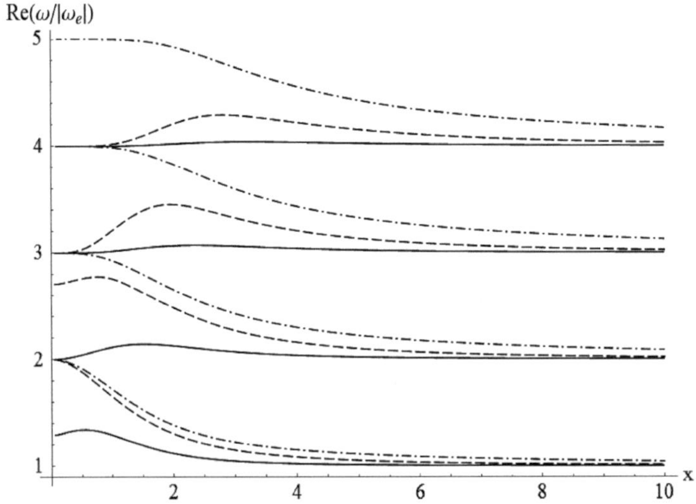


**Fig. 3.7** Eigenfrequencies of electron SCTM-modes versus normalized wavenumber $x = k_1\rho_e$. Solid, dashed and dash-dotted curves relate to the cases $Z = \Omega_e^2\omega_e^{-2} = 1.5$, 10 and 100, respectively

of these modes, the influence of increasing value of the external static magnetic field causes the approach of these curves to the corresponding electron cyclotron harmonics more and more. Therefore, the wavenumber range where the inequality $h_e > v_{Te}^2/c^2$ is not satisfied increases in this case. This means violation of the basic starting assumption that transverse relativistic effects can be neglected in solving the kinetic Vlasov-Boltzmann equation.

For the studied model of a metal waveguide with plasma filling, there are specific values of the electron SCTM-mode wavenumbers $k_1$, for which the group velocity of the corresponding mode is equal to zero. This value of $k_1$ is different for each electron cyclotron harmonic for the same set of other parameters of the considered waveguide. This peculiarity was not observed in the case of a dielectric waveguide [18], for which the group velocity of the electron SCTM-modes tends to zero in the approximation of very short ($k_1\rho_e \gg 1$) and very long ($k_1\rho_e \ll 1$) wavelengths, respectively, but never exactly is zero. The existence of such regions in the dispersion curves (eigenfrequency of these modes vs their wavenumber) where electron SCTM-modes cannot transfer their power could be applied for realization of local plasma heating regimes due to absorption of these waves.

Now the effect of transverse inhomogeneity of the plasma particle density, which fills the considered metal waveguide structure, on the dispersion properties of the SCTM-modes is considered. This analysis can be performed only numerically. To do that it is assumed that the plasma consists of a sequence of homogeneous plasma layers of different thickness and plasma density [22].

At first the model of a plasma filled waveguide is presented (see Fig. 3.8). The region $0 < z < a_1$ is the transition plasma layer with thickness $a_1$ and plasma

**Fig. 3.8** Schematic of the model of non-uniform plasma filling with one transient layer

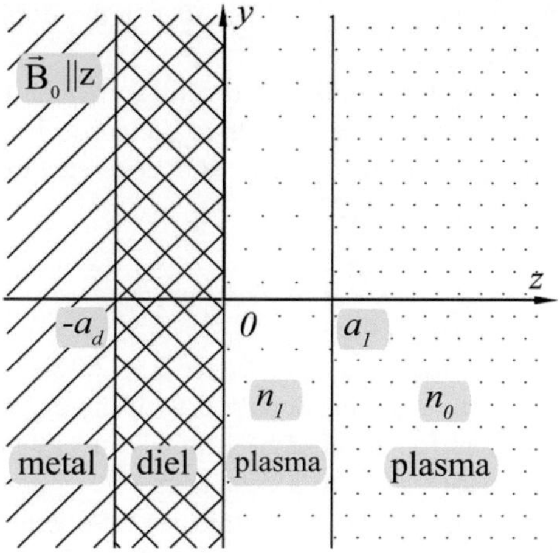


density $n_1$. The main region of the plasma with a uniform density $n_0$ occupies the half-space $z > a_1$. From the metal wall of the studied waveguide the plasma filling is separated by a dielectric layer, whose thickness is $a_{\mathrm{d}}$ and its dielectric permittivity is $\varepsilon_{\mathrm{d}}$. To find the dispersion relation of SCTM-modes propagating in such a waveguide structure one can apply the boundary conditions, which have been already discussed in Sect. 3.1 for the case of a semibounded plasma and at the beginning of the present Sect. 3.3 for a plasma layer with restricted thickness. Enlarging the quantity of applied boundary conditions allows one to make the preliminary conclusion that the form of this dispersion relation is more complicated as compared with the case of a uniform plasma filled waveguide.

Studying the dispersion properties of the electron SCTM-modes propagating in non-uniform plasma, one applies the assumptions made at the beginning of this chapter. The unperturbed distribution function of the plasma particles is a Maxwellian, the reflection of plasma particles from the plasma-dielectric and plasma-plasma interfaces is assumed to be described by the model of mirror interaction, the studied surface cyclotron waves are assumed to be slow waves, and the plasma spatial dispersion along the direction of the external static magnetic field is assumed to be weak. The last assumption allows one to apply the expression of the dielectric permittivity tensor that was obtained in Sect. 2.2 for description of electromagnetic features of the plasma together with (3.1) and (3.23) which describe the field of electron SCTM-modes in both plasma regions with different plasma density. The set of equations (3.23) describes the electromagnetic field of these SCTM-modes also in the dielectric layer separating the plasma filling from the metal wall of the waveguide if one makes there the following replacements: $\varepsilon_{11} \to \varepsilon_{\mathrm{d}}$, $\varepsilon_{33} \to \varepsilon_{\mathrm{d}}$, $\varepsilon_{13} \to 0$.

Since the expressions for the impedances of the dielectric region and the plasma layer region are calculated in this section [see expressions (3.24)] and the impedance of a semibounded plasma was calculated in Sect. 3.1, the derivation of the dispersion equation for SCTM-modes in the studied waveguide structure should not cause difficulties.

Using the mentioned formulas for the impedances of the plasma regions and the impedance of the protective dielectric layer, one can obtain the following dispersion relation for electron SCTM-modes in the case of the chosen model of plasma inhomogeneity [22]:

$$\sqrt{\varepsilon_{11}(n_0)(\varepsilon_{33}(n_0)+2B(n_0))} = f(n_1) + \varepsilon_{\rm d}\coth(|k_1|a_{\rm d}), \qquad (3.29)$$

where $f(n_1) = \varepsilon_{11}(n_1)(\varepsilon_{33}(n_1)+2B(n_1)) \cdot \tanh(|k_1|\Theta(n_1)a_1)$.

Comparing this equation with the dispersion relation (3.9) of the SCTM-modes obtained for the case of a metal waveguide with a uniform plasma filling, one can see that (3.29) has a new term that appeared due to the chosen model of plasma inhomogeneity, namely uniform semibounded plasma with a transition plasma layer. The value of this term is determined by the parameters of this transition plasma layer. This change in the radial plasma density profile leads to the existence of two solutions of the dispersion relation (3.29) in contrast to the previous case of uniform plasma [see (3.9)]. This means that in a plane waveguide structure filled with non-uniform plasma two surface cyclotron modes with TM polarization can propagate simultaneously. One of these modes exists in the case of a uniform plasma filling of the studied waveguide as well. This mode is appropriate to be called as the principal mode, and its frequency in the limiting cases of long and short wavelengths is given by the expressions (3.10)–(3.13). The other electron cyclotron mode exists according to the choice of the model of inhomogeneity of the plasma filling in this waveguide, or in other words, it can propagate there due to the presence of the plasma transition layer with a sharp boundary in the applied model of the plasma non-uniformity. This mode, which propagates along the interface between the plasma layer with density value $n_1$ and the main area of the plasma with the density $n_0$, is called the minor mode.

Of course, in real experimental practice, both modes can exist simultaneously only in the case of a layered semiconductor plasma waveguide structure or meta-plasma waveguide structure when there is a sharp boundary between two layers of solid state plasma-like materials with different concentration of charge carriers. In the case of a gaseous plasma, which is magnetically confined in the waveguide where it is impossible to have a sharp boundary between two regions of the plasma filling with different density, it should be clearly understood that no minor mode exists. That is why it is evident that this model of stepped plasma inhomogeneity cannot be applied to simulate the case of gaseous plasmas with arbitrary density profile. The correct description of the problem should be performed by application of the model with a large number of different plasma layers. Then the corresponding dispersion relation will have a complicated solution consisting of a large number of similar minor modes, which will characterize the propagation of electromagnetic electron cyclotron eigenwaves along the corresponding interfaces of all the modelled plasma layers, and the only single solution will describe the principal electron SCTM-mode.

Therefore, to approach the present step model to the description of an inhomogeneous gaseous plasma confined in a real laboratory device, one should apply a sequence of a large number of uniform plasma layers with appropriate densities of the charged plasma particles and corresponding values of thickness. Then instead of the dispersion relation (3.29) one will obtain a more complicated equation that can be written as follows:

$$\sqrt{\varepsilon_{11}(n_0)(\varepsilon_{33}(n_0)+2B(n_0))} = \sum_{i=1}^{N} f(n_i) + \varepsilon_{\mathrm{d}} \coth(|k_1|a_{\mathrm{d}}). \tag{3.30}$$

This equation has $N + 1$ solutions that describe the surface cyclotron waves with TM polarization propagating along a single plasma-dielectric boundary and $N$ boundaries between plasma layers with different thickness and density. An arbitrary non-uniform plasma density profile that is simulated by a step function leads to broadening the range of eigenfrequencies of SCTM-modes. In the case of such modelling the plasma non-uniformity, only numerical analysis will allow one to investigate the dependence of the real frequency of SCTM-modes on their wavelength in a wide range of parameters of the plasma filled waveguide.

Nevertheless, it should be emphasized that modeling of the plasma non-uniformity by a single transition layer has a methodological significance for studying the influence of non-uniformity of a gaseous plasma on cyclotron waves at all. Considering the method applied for finding the requested solution in the case of a single transition layer as an example, one can simply understand how to find exactly the solution that corresponds to a wave propagating just along the plasma-dielectric interface among a large number of solutions related to minor modes, which do not have physical meaning in the case of a gaseous plasma. In the following the dispersion relation (3.29) is investigated in details. In this expression the thickness of the transition layer of the plasma and the ratio of plasma densities in the region of main plasma and transition layer $n_0/n_1$ can be variables. The plasma density profile can also be considered as increasing or decreasing in perpendicular direction to the interface between dielectric layer and metal wall of the waveguide.

The obtained results are plotted in Figs. 3.9 and 3.10 for the first three electron cyclotron harmonics. There one can see the dependence of the studied electron SCTM-mode frequency, normalized to the absolute value of the fundamental electron cyclotron frequency, on the wavenumber, which is normalized to the electron Larmor radius. The parameters of the studied waveguide structure are as follows: $\Omega_e^2(n_0) = 10^3\omega_e^2$, $\varepsilon_{\mathrm{d}} = 1$ and $a_{\mathrm{d}} = 10^{-4}\rho_e$. The curves, which are labelled with triangles, correspond to the principal SCTM-mode (the mode, which propagates along the plasma-dielectric interface). The snowflake marked dispersion curves are related to the minor modes, for cases $a_1 = 10^{-4}\rho_e$ (solid) and $a_1 = 10^{-6}\rho_e$ (dashed), respectively. To estimate the influence of the thickness of the transition plasma layer on the dispersion of minor modes one can compare the shapes of these curves. The abscissa axis has a logarithmic scale. As a result of the numerical analysis it is found that the

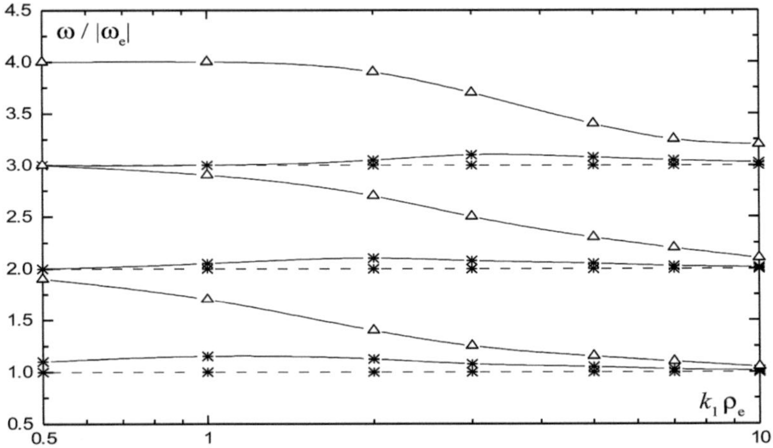


**Fig. 3.9** Dependence of the electron SCTM-mode frequencies on the wavenumber in the case of a transition plasma layer with the density $n_1 = 0.1n_0$

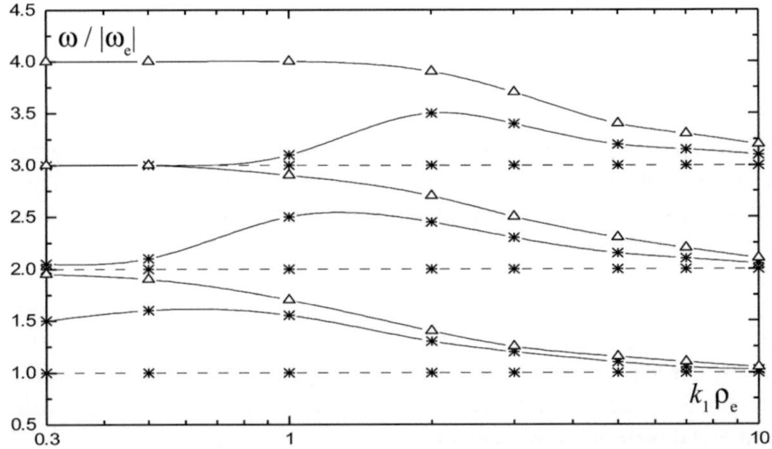


**Fig. 3.10** Dependence of the electron SCTM-mode frequencies on the wavenumber in the case of a transition plasma layer with the density $n_1 = 0.3n_0$

frequency of the principal electron SCTM-mode is independent of the parameters of the transition layer, in contrast to the frequencies of the minor modes.

With increasing number $s$ of electron cyclotron harmonic, as well as with decreasing thickness of the transition plasma layer and increasing relative value $n_0/n_1$ the conditions of existence of the minor modes deteriorate. This means that the electron SCTM-mode frequency becomes more and more closer to the corresponding resonance line: $s|\omega_e|$. In addition to that, with increasing number $s$ of electron cyclotron harmonic, the range of wavelengths, where the minor modes may propagate, becomes smaller and the range as a whole is shifted to the short-wavelength region. For decreasing thickness of the transition plasma layer $a_1/\rho_e$ and/or for increasing ratio $n_0/n_1$, the frequency of the minor mode approaches to the frequency of the principal electron SCTM-mode, and the problem reduces to the case of a waveguide with uniform plasma filling.

The case, when the plasma density profile is assumed to increase along the direction towards the waveguide wall (such situation can be realized in the cases of metaplasma filling or some technological process where the metal wall is utilized as element of an antenna system applied for excitation of waves in the waveguide), has been also examined for the first three cyclotron harmonics. It is found that if the density of charged particles in the transition layer plasma is much larger than their density in the main plasma region and if the thickness of this layer is much smaller than the size of the main plasma region, the frequencies of both modes will become the same. The difference between their frequencies increases with increasing electron cyclotron harmonic number. By varying the parameters of the transition layer of the plasma filling (density, transverse size), the frequency of minor modes is changing. In contrast to this, the frequency of the principal mode is constant. Therefore, this procedure allows one to determine the frequency of the principal electron SCTM-mode.

This means that modelling of a plasma inhomogeneity by many homogeneous plasma layers with different values of their particle density and thickness is an adequate tool for determining the frequency of electron SCTM-modes. To select the principle mode among the many solutions that arise when using this method one just has to slightly change the layer parameters, for example, their thickness. Then the frequencies of non-physical (minor) modes change, while the frequency of the principal electron SCTM-mode remains constant. To prove this statement a numerical analysis of the influence of the number of transition plasmas layers is carried out. The results of this analysis are presented in Fig. 3.11 for the case of SCTM-mode propagation at the second electron cyclotron harmonic.

Analyzing the shape of curves plotted in Fig. 3.11 one can make the following two conclusions: (i) increasing number of modeling transitional layers does not lead to an increasing difference between the frequency of these electron SCTM-modes calculated in the cases of uniform and non-uniform plasma filling, (ii) there is a definite limiting value of the frequency of electron SCTM-modes propagating along the boundary of a non-uniform gyrotropic plasma, which is calculated by the proposed modeling method. Thus, this theoretical method is appropriate for studying the influence of plasma non-uniformity on the propagation of such electromagnetic waves, whose frequency is only weakly dependent on the plasma density.

Numerical analysis of the influence of the plasma particle density profile on the dispersion of electron SCTM-modes has been performed for the three types of plasma density profiles, which are plotted in Fig. 3.12. Results of the analysis are shown in Fig. 3.13. It allows to make the following conclusions. First, an increasing density gradient leads to a decrease of the frequency of the studied SCTM-modes. Second, an increasing density gradient is the reason for the shift of the value of their wavenumber, where the group velocity of these modes is equal to zero, to the region of long wavelengths. The analysis of the curves depicted in Figs. 3.11 and 3.13 shows that

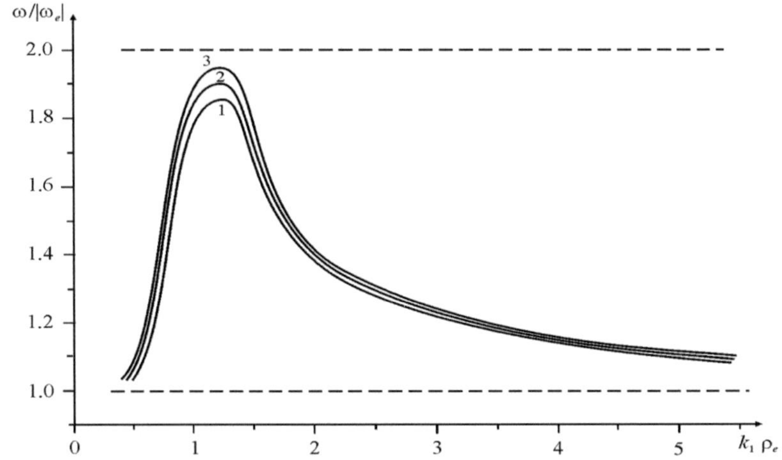


**Fig. 3.11** Illustration of the accuracy of modelling plasma non-uniformities by a sequence of transition plasma layers. The numbers 1, 2 and 3 denote the dispersion curves for the electron SCTM-mode at the second harmonic electron cyclotron resonance calculated in the cases of 5 and 50 transition layers and uniform plasma filling, respectively

plasma density non-uniformity leads first of all to some weak decrease of the electron SCTM-mode frequency. This is explained by the fact that the transfer from uniform plasma to non-uniform plasma can be considered as a decrease of the plasmas density, and therefore this leads to a decreasing value of the eigenfrequency of electron SCTM-modes (as it was shown in the Sect. 3.1). Since application of a small number of modeling transition plasma layers gives a smaller value of the SCTM-mode frequency as compared with the value, which is calculated for the case of a large quantity of modeling plasma layers, there is an optimal quantity of these layers.

In this section it is proved that a non-uniformity of the plasma filling of a metal waveguide leads to a decrease of the electron SCTM-mode frequency as compared with the case of uniform plasma filling. The value of this frequency reduction depends on the thickness of the plasma transition layer, the plasma density gradient and the number of the cyclotron harmonic. The example of simulation of the plasma non-uniformity by a transition layer has shown in which way one can select the solution that corresponds to waves propagating along the interface between the non-uniform plasma and the dielectric coating of the waveguide. Here it is proved by numerical analysis that modeling of a plasma density non-uniformity by a sequence of plasmas layers, which have different particle density and thickness, is quite acceptable for description of SCTM-mode propagation in a metal waveguide partially filled with non-uniform plasma. Moreover, even modeling of the plasma density non-uniformity by a very simple model, namely by just the main plasma region and a single transition plasma layer, allows one to obtain a good numerical agreement compared with the case of application a model with a large number of transition plasma layers. It is proved that the influence of the ratio of the plasma density calculated in the peripheral

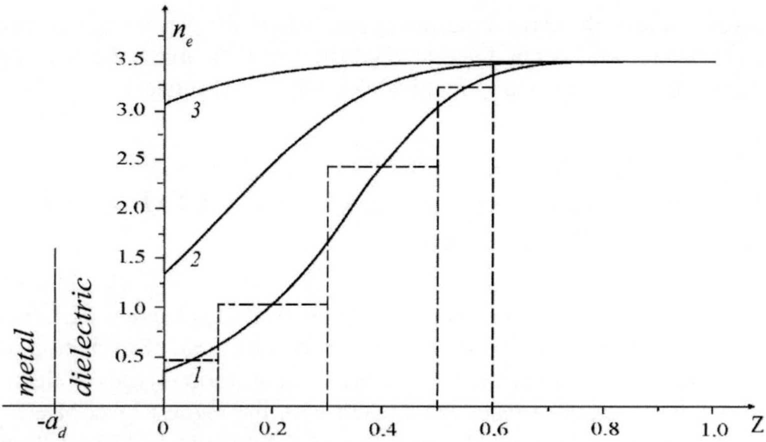


**Fig. 3.12** Schematic of the applied model of non-uniform plasma filling of the waveguide. The solid lines show different types of peripheral plasma density (in units of $10^{-18}$ $m^{-3}$). The curve marked by number 1 corresponds to the plasma density profile, which was planned to be realized in the stellarator W7-X [23]. The dashed lines indicate the same plasma density profile modelled by different layers of uniform plasma. The curves marked by 2 and 3 describe plasma density profiles which are typical for the cases of some plasma technology processes

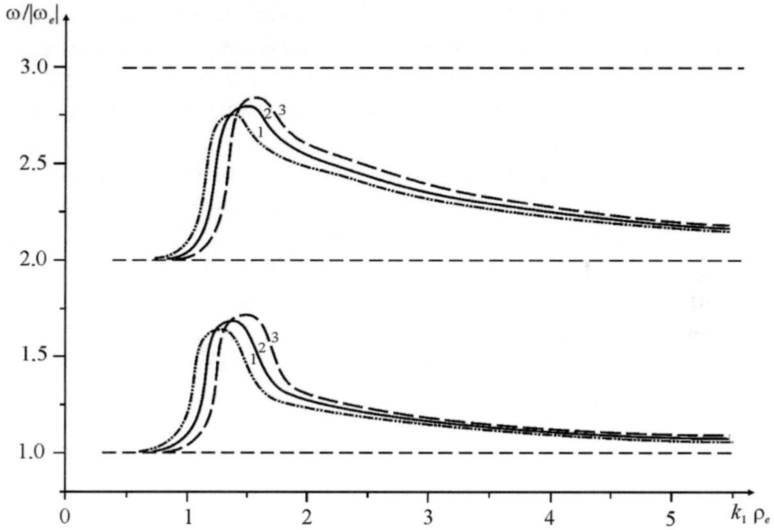


**Fig. 3.13** Dispersion curves of electron SCTM-modes calculated for the plasma density profiles plotted in Fig. 3.12. The numbers 1, 2 and 3 correspond to the same plasma density profiles as presented there

plasma region and in the central plasma region (where the plasma density profile is assumed to be uniform) is not essential compared with the influence of a thickness of the transition layer on the amplitude of electron SCTM-modes.

## 3.4 Excitation of Surface Electron Cyclotron TM-Modes by Charged Particle Beams

The complete study of electromagnetic wave propagation processes in plasma filled devices has to provide information about the important issue of excitation of these waves in different types of plasma waveguides. Among various types of electromagnetic waves, just the eigenmodes are excited with the highest level of efficiency. Utilization of a charged particle beam is one of the most widely applied methods of excitation of plasma filled waveguide structures. The plasma-beam instability was discovered more than fifty years ago [24, 25], but until the present time there are some questions regarding the interaction of a charged particle flow with plasma waves, which remain unsolved completely. This is especially true in the case of the interaction between a charged particle beam and eigenmodes of a bounded plasma that is confined in different types of plasma traps. Excitation of surface type waves essentially distinguishes from the case of bulk wave excitation. Therefore, the study of beam-plasma instabilities in restricted plasma structures is the subject of intense theoretical and experimental studies until the present time. For example, various aspects of excitation of surface waves in a bounded plasma are studied in the monographs [26–31] (see also references therein).

This section is devoted to the theoretical study of the problem of beam excitation of surface electron cyclotron TM-modes. A model of a semibounded plasma with the following geometry is considered (see Fig. 3.14). The plasma region occupies the space $z \geq 0$. A cold charged particle beam moves with the velocity $\vec{u}$ in the region $z < 0$, along the $\vec{x}$ axis. Thus, the beam velocity is oriented along the wave vector of the SCTM-modes and across the applied external static magnetic field $\vec{B}_0$.

In the following, the proposed model of the charged particle beam is discussed in more detail. It is well-known that in an external magnetic field charged particles move along complicated trajectories, but under certain circumstances, this trajectory can be considered roughly as a straight line. At first, for example, if the Larmor radius of the gyrating motion of the beam particles is much larger than the characteristic spatial scales of the considered problem such as the wavelength of the excited electromagnetic wave, or the distance, which the beam propagates during its linear interaction with the excited wave. At second, for instance, if in addition to the static magnetic field $\vec{B}_0 \| \vec{z}$ a static electric field $\vec{E}_0 \| \vec{y}$ is present in the area, where the charged particle beam moves (in other words if there is motion of the beam in crossed electric and magnetic fields, namely along the $\vec{x}$ axis). In the first case, the beam particles move along a circle with the Larmor radius $R_L = u/|\omega_b|$, where $\omega_b = e_b B_0/(m_b c)$ is the cyclotron frequency of the beam particles. The radius of curvature $R_L$ of the

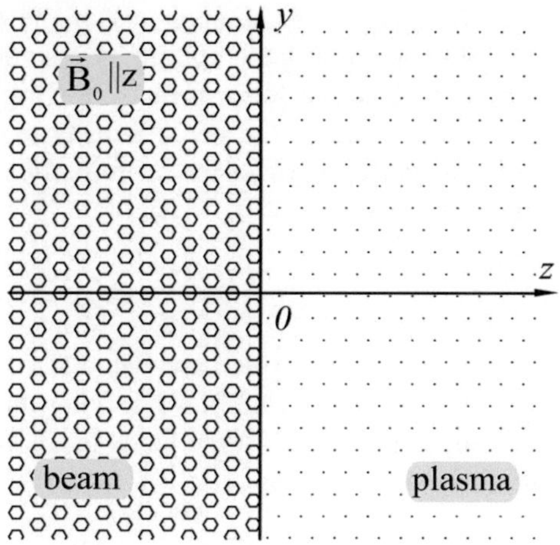


**Fig. 3.14** Schematic of the problem "excitation of SCTM-modes by charged particle beams in semibounded plasmas"

trajectory, along which the charged particle motion is considered, should be much larger than the wavelength $\lambda$ of the studied SCTM-mode. The wavelength can be estimated from the condition of the Cherenkov resonance $\omega \approx k_1 u \approx s|\omega_\alpha|$, where $s$ is the number of the cyclotron harmonic and the index $\alpha$ indicates the type of the charged plasma particle ($\alpha = e$ for electrons and $\alpha = i$ for ions, respectively). Then the value of the SCTM-mode wavelength is given by $\lambda \approx u/|s\omega_\alpha|$. Therefore the condition for sufficient accuracy of the description of the propagation of the beam particles by rectilinear motion can be described by the following inequality: $R_L \gg \lambda$, or $u/|\omega_b| \gg u/|s\omega_\alpha|$, which is equivalent to $|s\omega_\alpha| \gg \omega_b$. This requirement can be easily fulfilled in the case of electron SCTM-mode excitation due to interaction with an ion beam, because these conditions require the inequality $|\omega_b/\omega| \sim m_e/m_i \ll 1$, which is satisfied with a considerable margin. However, also in the case of ion SCTM-modes, of course, one can utilize a "heavy-ion beam", where the mass of the beam particles significantly exceeds the mass of plasma ions: $m_b \gg m_i$.

If the motion of the charged particle beam is performed in crossed static electric and magnetic fields, in the framework of fluid theory the rectilinear beam motion can be realized under the condition that the Larmor rotational motion is negligible as compared to the drift motion of particles with the velocity $u = \upsilon_d = cE_0/B_0$.

In the present case the motion of these particles is described by the equations of magnetohydrodynamics where in the equation of particle motion the particle velocity is represented as the sum of the equilibrium value of the velocity and a small correction $\vec{\upsilon} = \vec{u} + \vec{\tilde{\upsilon}}$. Then one can find the following expressions for the values of the corresponding corrections to the components of the charged particle beam velocity $\tilde{\upsilon}_j$:

$$\begin{aligned} \tilde{\upsilon}_x &= i\omega_1 e_b E_x/[m_b(\omega_1^2 - \omega_b^2)], \\ \tilde{\upsilon}_y &= \omega_b e_b E_x/[m_b(\omega_1^2 - \omega_b^2)], \\ \tilde{\upsilon}_z &= i e_b E_z/(m_b \omega_1), \end{aligned} \tag{3.31}$$

where $E_x$ and $E_z$ are components of the electric field of the SCTM-modes, and $\omega_1 = \omega - k_1 u$. Using the continuity equation, where the density of beam particles is represented in the form of the following sum: $n_b = n_{0b} + \tilde{n}_b$, where $n_{0b}$ is the equilibrium value of the beam density and $\tilde{n}_b$ is a small correction (or in other words, the perturbation of the beam particle density), respectively, one can find the expression for $\tilde{n}_b$:

$$\tilde{n}_b = n_{0b}(k_1 \tilde{\upsilon}_x - i\partial \tilde{\upsilon}_z/\partial z)/\omega_1. \tag{3.32}$$

Using the values, which are found for the perturbed velocity (3.31) and the perturbed beam density (3.32), for solving the Maxwell equations one can derive the following expressions for the fields of SCTM-modes in the space region, where the particle beam is moving ($z < 0$):

$$\begin{aligned} ik_1 H_y &= ik\varepsilon_3^{(b)} E_z, \\ ikH_y &= -ik_1 E_z + \partial E_x/\partial z, \\ \partial H_y/\partial z &= ik\varepsilon_1^{(b)} E_x - k\partial(U_b E_z)/\partial z, \end{aligned} \tag{3.33}$$

where $\varepsilon_1^{(b)} = 1 - \frac{\Omega_b^2}{\omega_1^2 - \omega_b^2}$, $\varepsilon_3^{(b)} = 1 - \frac{\Omega_b^2}{\omega\omega_1}$, $U_b = u\Omega_b^2/(\omega\omega_1^2)$, $\Omega_b = \sqrt{\frac{4\pi\, e_b^2 n_{0b}}{m_b}}$ is the plasma frequency of the beam particles.

To derive the dispersion relation for describing the initial stage of the excitation of SCTM-modes as a result of their interaction with the charged particle flow moving over the plasma surface, one can apply the following boundary conditions for the tangential components of the electric and magnetic SCTM-mode fields. These conditions can be formulated by integration of the set of equations (3.33):

$$\begin{aligned} E_x(z = +0) &= E_x(z = -0), \\ H_y(z = +0) &= H_y(z = -0) + kU_b E_z(z = -0). \end{aligned} \tag{3.34}$$

Solving the Maxwell equations in the region which is filled by the particle beam ($z < 0$) in the approximation of slow waves under the following condition for the phase velocity of SCTM-modes $\omega/k_1 \ll c$, which can be rewritten as $k \ll k_1$, one can obtain a second order differential equation for the magnetic field of these modes. Its solution, under the fulfillment of the boundary condition to have a finite value in the limiting case $z \to -\infty$, has the following form: $H_y \propto \exp(\kappa_b z)$, where the parameter is $\kappa_b = \sqrt{\varepsilon_1^{(b)} k_1^2/(\varepsilon_3^{(b)} - k_1 U_b)}$. The electrical components of the SCTM-mode fields have the same dependence on the transverse coordinate $Z$ and therefore they can be expressed in terms of the tangential magnetic field mentioned above. Using this statement, one can calculate the expression for the normal component of electric field on the plasma interface $E_z(-0)$ from the set of equations (3.33):

$$E_z(-0) = -k_1 H_y(-0)/(k\varepsilon_3^{(b)}). \tag{3.35}$$

In Sect. 2.1 for the model of semibounded plasmas, the SCTM-mode impedance $Z_{pl}(0) = E_x(z = +0)/H_y(z = +0)$ has been calculated on the plasma interface $z = +0$ [see (3.8)]. Using the boundary conditions (3.34), the expression for the normal electric field of these modes on the plasma interface (3.35) and the expression for the plasma impedance (3.8) one can derive the equation describing the beam excitation of electron SCTM-modes for semibounded waveguide plasmas:

$$\varepsilon_{11}(\varepsilon_{33} + 2B) = \varepsilon_1^{(b)}(\varepsilon_3^{(b)} - k_1 U_b). \tag{3.36}$$

It should be recalled that here on the left-hand side of equation (3.36), which earlier has been denoted as $D_0(\omega, k_1)$, only the Hermitian parts of the respective components of the plasma dielectric permittivity tensor $\varepsilon_{ij}$ are written. This is explained by the fact that anti-Hermitian parts of the tensor $\varepsilon_{ij}$ describe the kinetic damping of these modes that is out of scope of our consideration in this section.

The existence of the beam terms on the right-hand side of equation (3.36) under definite resonance conditions results in the appearance of solutions expressed in complex values that describe the beam excitation of electron SCTM-modes. Its left-hand side $D_0(\omega, k_1)$ can be decomposed into a Taylor series in the vicinity of their eigenfrequencies $\omega = \omega_0 + \gamma$, where $\omega_0$ is the eigenfrequency of the dispersion relation of these modes if there is no particle beam, and $\gamma$ is the frequency correction, which is caused by the presence of the beam. Its value is small: $|\gamma| \ll \omega_0$. The analysis of the dispersion relation (3.9) has shown that the values of the diagonal components of the plasma dielectric permittivity tensor are $\varepsilon_{11}(\omega_0) \approx 0$ and $|\varepsilon_{33} + 2B| \gg 1$. That is why the product of the factors on the left-hand side of equation (3.36) for the theoretical analysis of SCTM-mode excitation by a charged particle beam should be represented in the following form:

$$D_0(\omega, k_1) \approx (\gamma + i\nu)(\varepsilon_{33} + 2B)\Big|_{\omega_0} \cdot \frac{\partial \varepsilon_{11}}{\partial \omega}\Big|_{\omega_0}, \tag{3.37}$$

where $\nu$ is the effective collision frequency between the plasma particles.

First, the case of beam instabilities, when the inequality $|\gamma| \gg \nu$ is satisfied, is considered. This means that the correction of the SCTM-mode eigenfrequency is larger than the collisional frequency. Under these conditions, the kinetic energy of the particle beam is directly spent to amplify the amplitude of the eigenwaves in the plasma filled waveguide. In the other case, when the opposite inequality $|\gamma| \ll \nu$ is realized, the instability is called as dissipative instability, which is characterized by a growth rate that explicitly depends on the frequency of collisions between the plasma particles. As it is shown in Sect. 3.3, the damping process caused by particle collisions is the main channel of energy dissipation of electron SCTM-modes. Since the oscillating system, consisting of the plasma and the charged particle beam, is characterized by a "negative" energy, the reduction of energy of this oscillating system is precisely transferred into an increase of the amplitude of the excited modes. However, one can see that the growth rates of the dissipative instabilities are smaller than the growth rates of the beam instabilities.

Under the condition of a low density of the charged particle beam as compared with the plasma particle density, which means $\Omega_b^2 \ll \Omega_\alpha^2$, the right-hand side of equation (3.36), which can be denoted as $D_b$ for obtaining a convenient writing form, can be transformed into a simplified form for studying different types of instabilities and different types of resonant conditions. For example, if the condition of Cherenkov resonance $\omega_0 \approx k_1 u$, or in other form $\omega = \omega_0 + \gamma = k_1 u + \gamma$ is realized, one can find the following approximate expression for the right-hand side of equation (3.36):

$$D_b \approx -\Omega_b^2/\gamma^2. \tag{3.38}$$

If the resonance condition for the anomalous Doppler effect $\omega_0 \approx k_1 u - |\omega_b|$ is satisfied, the expression for $D_b$ can be approximately rewritten in the following way:

$$D_b \approx \Omega_b^2/(2\gamma|\omega_b|). \tag{3.39}$$

An analytical solution of equation (3.36) can be obtained in the limiting cases of weak ($y_\alpha \ll 1$) and strong ($y_\alpha \gg 1$) spatial dispersion along the plasma interface. In these limiting cases one can transfer (3.36) to the form:

$$\gamma + i\nu \approx -D_b \cdot f(y_\alpha, s), \tag{3.40}$$

where, under the condition of validity of inequality $y_\alpha \gg 1$, the function $f(y_\alpha, s)$ is given by:

$$f(y_\alpha \gg 1, s) \approx \omega_0^3 \Omega_\alpha^4/(4\pi\, y_\alpha^4 \omega_\alpha^6). \tag{3.41}$$

Under the condition of weak plasma spatial dispersion ($y_\alpha \ll 1$) the function $f(y_\alpha, s)$ can be written in the following form:

$$f(y_e \ll 1, s) \approx \frac{(s^2-1)^4 I_s^2(y_e) y_e \omega_\alpha^2 \omega_0^3 \Omega_\alpha^{-4}}{4 I_1^2(y_e)[8 I_1^2(y_e) + (s^2-1)^2 I_s(y_e)]}. \tag{3.42}$$

After that one easily can find analytical expressions for the growth rates of the instabilities of electron SCTM-modes caused by their interaction with charged particle beams.

The expressions for the growth rates of SCTM-mode instabilities, which are solutions of equation (3.36), are derived now in the case of beam instabilities. Under the condition of Cherenkov resonance, the expression for the electron SCTM-mode growth rate is as follows:

$$\mathrm{Im}(\gamma) \approx \frac{\sqrt{3}}{2} \frac{\left[\Omega_b^2 f(y_\alpha, s)\right]^{1/3}}{|s\omega_\alpha|} \omega_0. \tag{3.43}$$

On the other hand, under the condition of the anomalous Doppler effect, the value of this growth rate becomes smaller:

$$\mathrm{Im}(\gamma) \approx \omega_0\left[\Omega_b^2 f(y_\alpha, s)/(2s^2\omega_\alpha^2|\omega_b|)\right]^{1/2}. \tag{3.44}$$

If the conditions of a dissipative instability are realized, the value of the electron SCTM-mode growth rate becomes smaller once again compared with the case of resonant beam-plasma instability. For example, in the case of Cherenkov resonance one can derive the following expression:

$$\mathrm{Im}(\gamma) \approx \omega_0\left[\Omega_b^2 f(y_\alpha, s)/(2\nu\, s^2\omega_\alpha^2)\right]^{1/2}. \tag{3.45}$$

In the case of anomalous Doppler effect, the expression for the growth rate of electron SCTM-modes has the following form:

$$\mathrm{Im}(\gamma) \approx \omega_0 \frac{\Omega_b^2 f(y_\alpha, s)}{2\nu|s\omega_\alpha\omega_b|}. \tag{3.46}$$

Taking into account those technical difficulties, which have to be solved in the case of confining a gaseous plasma by an external static magnetic field oriented perpendicularly to the plasmas interface, the parameters of an experimental device, which could be utilized for beam excitation of electron SCTM-modes in the case of solid state plasmas are estimated in [30, 31]. It could be either semiconductor plasmas [28], or a meta-plasma medium [32]. As example, a semiconductor plasma with a concentration of conducting electrons in the conduction band of ~$10^{14}$ cm$^{-3}$ is considered. The effective electron mass is about $10^{-28}$–$10^{-29}$ g, the temperature of the medium is ~100 K, and the value of the applied external static magnetic field is assumed to be $B_0 \approx 0.1$ T. If a proton beam with the particle density ~$10^8$ cm$^{-3}$ is utilized, then, in order to excite electron SCTM-modes, one needs a proton accelerator with a beam energy of $W = m_b u^2/2 = 30-300$ keV and a current density of ~$10^{-4}$–$10^{-3}$ A/cm$^2$. The linear size of the active region of the semiconductor should be in the order of $L$ ~15–40 cm. This value has been determined by the condition that the linear growth rate of electron SCTM-modes can be larger than the value of the parameter $u/L$.

Thus under the condition of weak spatial dispersion of the semiconductor plasma medium ($y_e = 0.1$) the growth rate of the beam instability of the studied modes according to expression (3.43) has a magnitude of about $7 \times 10^5$–$10^7$ s$^{-1}$. For the range of short wavelengths (which means the satisfaction of the condition of strong spatial plasma dispersion: $y_e \gg 1$) the value of the electron SCTM-mode growth rate can be slightly increased. Theoretical estimations show that in this case the growth rate of these modes could be about $\sim 3 \times 10^7$ s$^{-1}$. Thus, it is by three orders smaller than the electron cyclotron frequency at the external static magnetic field $B_0$ mentioned above. However, one should take into account the circumstance that these estimations have been carried out just for the case of the model of semibounded

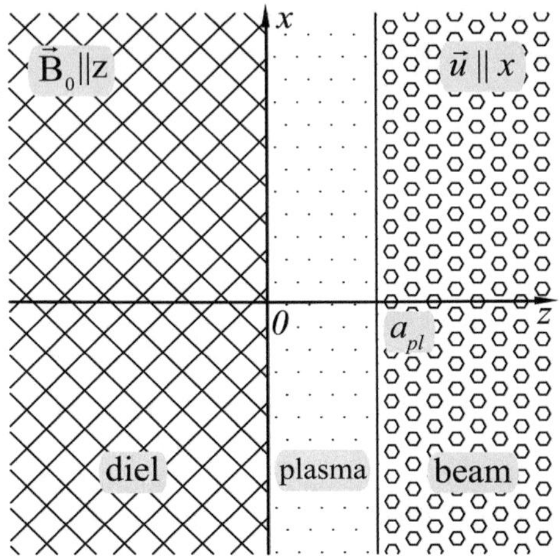


**Fig. 3.15** Schematic of the problem "excitation of SCTM-modes by charged particle beams in plasma layer"

plasmas. It is clear from the materials presented in Sect. 3.3 that if one takes into account a finite value of the transverse size of the plasma-like medium the growth rate of electron SCTM-mode excitation increases due to their interaction with the charged particle beam.

Now, a simple model of the dielectric waveguide is considered (see Fig. 3.15). It is filled with a semiconductor or meta-plasma layer, whose boundaries have the following coordinates $z = 0$ and $z = a_{\text{pl}}$. The external static magnetic field has the orientation $\vec{B}_0 \| \vec{z}$. The dielectric permittivity of the waveguide wall is approximately equal to unit. A cold flow of charged particles is moving above the interface of this plasma layer (in the region $z > a_{\text{pl}}$) parallel to the $\vec{x}$ axis. The dispersion properties of electron SCTM-modes propagating in plasma filled waveguides with this geometry for the case, when there is no charged particle beam, are studied in Sect. 3.3. There it is shown that a restriction of the transverse size of the plasma region leads only to the appearance of a trigonometric factor in the corresponding dispersion relation as compared with the case of a semibounded plasma. That is why one can write down the equation describing the excitation of electron SCTM-modes in the case of the waveguide filled with a plasma layer without detailed preliminary study. Doing that it should be underlined that the boundary conditions (3.34) in the case of the waveguide filled with a restricted plasma layer is valid if one replaces the location of the plane along which the beam is moving. Therefore, in this case it is the plane $z = a_{\text{pl}}$, but not the plane $z = 0$, as it was for the model of semibounded plasmas. As a result, one can derive the following equation instead of (3.36):

$$(\varepsilon_{33} + 2B)\varepsilon_{11} \tanh^2[|k_1|\, a_{\text{pl}}\sqrt{\varepsilon_{11}/(\varepsilon_{33} + 2B)}] = D_b, \tag{3.47}$$

where the expression of the right-hand side $D_b$ is the same as in (3.36).

Equation (3.47) can be studied analytically for the following two limiting cases. If the plasma layer is sufficiently thick, so that the inequality

$|k_1|\, a_{\rm pl}\sqrt{\varepsilon_{11}/(\varepsilon_{33}+2B)} \gg 1$ is valid, (3.47) transfers asymptotically into (3.36). Therefore, in the following, the other limiting case, when the thickness of the plasma layer is sufficiently thin and the inequality $|k_1|\, a_{\rm pl}\sqrt{\varepsilon_{11}/(\varepsilon_{33}+2B)} \ll 1$ is valid, is considered. Then (3.47) can be written in the following form for the case of beam instability of SCTM-modes:

$$(\gamma + i\nu)a_{\rm pl}k_1 \frac{s^2\Omega_e^2 \exp(-y_e) I_s(y_e)}{\omega_0^3 y_e h_e} = D_b. \qquad (3.48)$$

It should be noted that in the case of ion SCTM-modes equation (3.48) can be also used, but of course one has to replace all indices "$e$" by the indices "$i$" (ion parameters). The accuracy of this substitution is determined by the neglected small terms, whose values are of the order $m_e/m_i$, as it has been shown previously in Sect. 3.1. Therefore, there is no fundamental mathematical difference between these two cases.

Since the beam particle density is supposed to be small compared to the plasma density, the right-hand side of equation (3.48) is essential only for the Cherenkov resonance condition or the anomalous Doppler effect.

In the case of Cherenkov resonance and validity of the equality $\omega_0 \approx k_1 u$, the growth rates of electron SCTM-modes excited by beams of charged particles, which are moving over the plasma interface, are as follows:

$$\mathrm{Im}(\gamma) \approx \omega_0 \xi \left[\frac{\Omega_b^2 y_e h_e^2 \exp(y_e)\psi}{\Omega_e^2 s^2 I_s(y_e) a k_1}\right]^p, \qquad (3.49)$$

where the factor $\xi$ has the two values: $\xi = \sqrt{3}/2$ or 0.5, the parameter $P$ can be equal to either 1/3 or 1/2, and the parameter $\psi$ is $\psi = 1$ or $\psi = \omega_0/\nu$ under the conditions of resonant beam instability and dissipative instability, respectively.

In the case of the anomalous Doppler effect (under the condition $\omega_0 \approx k_1 u - |\omega_b|$) electron SCTM-modes are excited with smaller values of the growth rate:

$$\mathrm{Im}(\gamma) \approx \omega_0 \cdot \left[\frac{\Omega_b^2 y_e h_e^2 \exp(y_e)\omega_0 \psi}{2\Omega_e^2 s^2 I_s(y_e) a k_1 |\omega_b|}\right]^p. \qquad (3.50)$$

From a mathematical point of view this decrease is connected with an increase of the parameter $p$. Its value is equal to either 0.5, or 1 for the cases of resonant beam and dissipative instabilities, respectively. The value of the parameter $\psi$ is the same as it was for Formula (3.49). From a physical point of view, it is an ordinary well-known fact that beam-plasma interaction happens in a slower way under the condition of anomalous Doppler effect as compared with the case of Cherenkov resonance.

To confirm the statement on the increasing value of the growth rate of the electron SCTM-mode instability under the condition of the model of plasma layer filling with restricted value of its thickness one can carry out numerical estimations. A semiconductor plasma with a concentration of conductivity electrons of ~$10^{14}$ cm$^{-3}$

is considered. It is immersed into an external static magnetic field with a weak strength ~0.1 $T$ at the temperature $T$ ~ 100 K. The thickness of the semiconductor layer is supposed to be $a_{\text{pl}}$ ~ 0.001 cm, and the effective mass of the conductivity electrons is ~$10^{-27}$ g. It is assumed that electron SCTM-modes at the second harmonic of the electron cyclotron frequency are excited by a proton beam (so mass of the beam particles is $m_b \approx 2\times10^{-24}$ g). Then under the condition of weak spatial plasma dispersion one can obtain an approximate expression for estimation of the maximum value of the growth rate using Formula (3.49)

$$\text{Im}(\gamma) \approx 10^7/(a_{\text{pl}}k_1)^{1/3}c^{-1}. \tag{3.51}$$

If one utilizes some artificial meta-plasma medium, which is characterized by an effective mass of the electric charge carriers that is less than that specified above, this will allow to slightly increase the electron SCTM-mode growth rate, but it will be a minor increase. In the case of strong spatial plasma dispersion ($y_e = 10$) for the same values of the other parameters of the studied beam-plasma system one can apply the following expression in order to estimate the maximum value of the SCTM-mode growth rate:

$$\text{Im}(\gamma) \approx 10^9/(a_{\text{pl}}k_1)^{1/3}c^{-1}. \tag{3.52}$$

If the thickness of the plasma-like layer is infinitesimally thin, the product $a_{\text{pl}}k_1$ is restricted from the side of small values only by the initial assumption of weak spatial plasma dispersion along the $\vec{z}$ axis. Thus one has to check only the inequality $k_3 \ll k_1$. Its upper limit is determined by the application of the model of thin plasma layer, namely by validity of the inequality $(a_{\text{pl}}k_1)^2\varepsilon_{11}/(\varepsilon_{33}+2B) \ll 1$. Therefore, one can pay attention to the validity of the following conditions for the thickness of the plasma layer in the expressions (3.51) and (3.52). In the case of weak spatial dispersion ($y_e \ll 1$) along the surface of the plasma one can find the inequality:

$$\frac{\omega_e^2 y_e}{\Omega_e^2 I_s(y_e)} \ll a_{\text{pl}}k_1 \ll \frac{2\Omega_e^2 y_e^2}{\omega^2 I_s(y_e)(s^2-1)^2}. \tag{3.53}$$

In the case of strong spatial dispersion ($y_e \gg 1$) these conditions are given by:

$$\sqrt{2\pi}\frac{\omega_e^2 y_e^{3/2}}{\Omega_e^2} \ll a_{\text{pl}}k_1 \ll \sqrt{8\pi}\frac{\omega_e^2 y_e^{5/2}}{s^2\Omega_e^2}. \tag{3.54}$$

Thus, although the variation of the transverse size of the plasma-like medium $a_{\text{pl}}k_1$ influences the form of the dispersion equation of the SCTM-modes only weakly, the growth rates of the excited modes strongly depend on the transverse plasma dimension. Even taking into account the restrictions (3.53) and (3.54) in the examples presented above, one can reduce the requested value of the beam energy to 700 eV, and nevertheless the growth rate of the electron SCTM-modes increases to

~$8 \times 10^6$ $s^{-1}$ for the limiting case of weak spatial dispersion ($y_e \ll 1$) and to about $4 \times 10^8$ $s^{-1}$ in the case of strong spatial dispersion ($y_e \gg 1$). Therefore, their values are larger by an order of magnitude than those which were obtained from the estimations for the model of semibounded plasmas. Consequently, the linear sizes of the active zone of semiconductor or meta-plasmas, obtained above for this model, are also overestimated by an order of magnitude. Thus, there are really samples of semiconductors (meta-plasmas) with useful length since the linear stage of beam-plasma instability can happen at the distance of several centimeters.

Generally speaking, this section is devoted to the development of the linear theory of electron SCTM-mode excitation by charged particle beams using a fluid approximation for description of particle motion. So the velocities of all beam particles were assumed to be the same at the initial stage of the studied instability. This means that the above mentioned resonance conditions were considered to be fulfilled for all particles of the beam. It is clear that in experimental practice, it is impossible to get such a perfect bunch of charged particles, so it should be noted that the actual values of the growth rates will be smaller than those obtained in this section.

Now the relationship between the phase velocity of the excited SCTM-modes and the velocity of the beam particles, when the conditions of Cherenkov resonance [see Formula (3.38)] or anomalous Doppler resonance (see Formula (3.39)) are satisfied, is estimated. From the analysis of equation (3.40) one can show that under the condition of Cherenkov resonance the cubic equation $\gamma^3 \approx A > 0$ for the frequency correction of these TM modes can be derived. Thus the real part of the frequency correction for the beam-plasma instability fulfills the condition $\mathrm{Re}\gamma < 0$. This means that the phase velocity of the excited SCTM-modes is lower than the beam velocity under the condition of Cherenkov resonance. Under condition of the anomalous Doppler effect, it is clear from the condition $\omega \approx k_1 u - |\omega_b|$ that the beam velocity exceeds the phase velocity of the electron SCTM-modes: $\omega / k_1 < u$. This inequality means that as a result of the studied instability, which develops in the beam-plasma system, a redistribution of the power in this system takes place. The charged particle beam decelerates and transfers its kinetic energy to the electromagnetic field of the excited slow wave, which is an eigenmode of the plasma system. Such slow waves, as it is shown for the cases of different types of excited waves [26–29], are waves with so-called "negative energy". The content of this term (negative energy) in the case of classical physics is absolutely different compared with the case of quantum physics; it means that growth of the slow wave amplitude happens just due to capturing of kinetic energy from the beam particles, which interact with this slow wave. In other words, propagation of a wave with negative energy in a beam-plasma structure means that the energy of the particle beam, which is not disturbed by the excited slow mode, is larger than the energy of the beam, which interacts with the slow mode. Nevertheless, one can understand that the total energy of the beam-plasma system has a positive value and it is composed of energy generated by waves and of the energy of the beam particles.

In this section, it is proved theoretically that the instability of slow SCTM-modes can be realized under different regimes, namely: resonant beam-plasma instability and dissipative instability. In the resonant case two different resonance excitation conditions can be satisfied, namely: Cherenkov and anomalous Doppler resonances. Regarding this, it should be emphasized that the beam-plasma instability, even under the condition of Cherenkov resonance, has a different physical mechanism than the well-known Vavilov-Cherenkov effect. Under the condition of the beam-plasma instability the charged particle beam, losing kinetic energy, decelerates, and breaks up into separate bunches, which are captured into potential wells of the electric field of the excited wave. One should remember also that the radiation of electromagnetic power, which is excited by charged particle beams propagating in a dielectric medium under the Vavilov-Cherenkov effect, does not depend on the mass of the charged beam particles, while the dependence of the growth rates of beam-plasma instability on the mass of the beam particles is essential. Moreover, this phenomenon is especially evident in the case of relativistic beams interacting with plasmas. In this case one has to take into account the so-called relativistic factor (Lorentz factor) that increases the effective mass of the beam particles compared with the value of their rest mass.

As it is observed for different beam-plasma instabilities, interaction under the condition of Cherenkov resonance is stronger (because, the right-hand side of equation (3.36) $D_b \propto \gamma^{-2}$), than in the case of the Doppler effect (no matter: either anomalous or normal), when $D_b \propto \gamma^{-1}$. Therefore, it is clear how the instability growth rates of electron SCTM-modes under these two resonant conditions correlate. The growth rate for Cherenkov resonance is larger than in the case of the anomalous Doppler effect.

It is also interesting to note that transfer of charged particle beam energy into excited modes under the conditions of Cherenkov and anomalous Doppler resonances is performed from the kinetic energy located in the longitudinal degree of freedom. This circumstance distinguishes both these cases from the beam-plasma instability under the condition of normal Doppler effect, when beam energy transfer happens from the both longitudinal and transverse (relatively to the direction of the static external magnetic field) degrees of freedom of the beam particles. Thus, for the case of beam-plasma instability under the condition of normal Doppler effect a reduction of longitudinal and transversal particle beam energy is observed. In contrast to this, under the condition of anomalous Doppler resonance, the transverse kinetic beam energy may be equal even to zero at the beginning of the interaction between plasma waves and the charged particles. However, it will increase during the process of beam-plasma interaction.

The losses of energy in the beam-plasma system, which can be caused by different dissipative processes, such as collisions between plasma particles, radiation of electromagnetic energy of waves propagating in the plasma etc., can under certain circumstances lead to dissipative instabilities. According to thorough studies (see e.g. [26, 29]), these instabilities develop in beam-plasma systems if the inequality $|\mathrm{Re}(\gamma \cdot \partial D_0/\partial\omega)| < |\mathrm{Im} D_0|$ is valid. For the case of a quite high plasma particle density this relation can be approximately replaced by the following inequality for the effective collision frequency of the plasma particles: $\nu \gg (n_b/n_{\mathrm{pl}})^{1/3}\Omega_e$. This inequality is almost never valid for the case of a non-bounded gas plasma. There the approach of a collisionless plasma is very good. But in the case of bounded plasma systems, when energy dissipation happens due to different processes, not only because of the collisions between the plasma particles, the effective collision frequency $\nu$, which models all the dissipative processes mentioned above, can have a quite large value. Usually dissipative phenomena are the reason for plasma relaxation to an equilibrium state. However, due to the presence of a charged particle flow, the power dissipation in the plasma can lead to instability of eigenmode oscillations of this system, as it is described above (Fig. 3.15).

## 3.5 Parametric Excitation of Surface Electron Cyclotron TM-Modes

The theory of parametric instabilities of electromagnetic waves in unbounded plasmas has been developed for a long time (see monograph [33] and references therein). That is why at the present time one may think that this theory is relatively well developed. Unfortunately, this statement is not true for the theory of parametric excitation of surface waves.

It should be pointed out, that the papers [33–35] present results of pioneer works devoted to the parametric excitation of bulk waves propagating in unbounded plasmas immersed into an external static magnetic field. There it was found that this excitation has essential distinguishing features as compared with the case of non-magnetized plasmas. The first papers devoted to parametric instabilities of bulk electron cyclotron waves in plasmas have been published more than forty years ago (see, e.g. [36, 37]). Unlike this theory of the bulk waves, the theory of surface waves at harmonics of the cyclotron frequency cannot show significant achievements, especially in the field of studying the parametric instabilities of these modes. That is why the main purpose of this section is to demonstrate the present situation in the area of developing the theory of electron SCTM-mode parametric instability. It is shown here, how one can examine the possibility of realization of the parametric excitation of these surface cyclotron waves at the harmonics of the electron cyclotron frequency, propagating in a metal waveguide with dielectric coating, by an external RF field. Studying the dependence of the growth rate of parametric instabilities of electron SCTM-modes on wavenumber, number of cyclotron harmonic, thickness of the plasma filling of the waveguide, and thickness and value of the dielectric permittivity of the metal wall coating of the waveguide is carried out also.

At first, the case when an external RF field influences the propagation of electron SCTM-modes in the waveguide structure is considered. The waveguide consists of a metal wall that has a protective dielectric coating of the thickness $a_{\mathrm{d}}$, the dielectric permittivity $\varepsilon_{\mathrm{d}}$, and a plasma filling, for which the model of semibounded medium is chosen (see Fig. 3.16). This choice is justified under the condition that the penetration depth of surface cyclotron waves into the plasma is much smaller than the transverse dimensions of the plasma. It is supposed as well that an external constant magnetic field $\vec{B}_0$ is directed along the $\vec{z}$ axis, so it is perpendicular to the plasma-dielectric interface. The external monochromatic pumping RF field $\vec{E}_0 \sin(\omega_0 t)$ is oriented

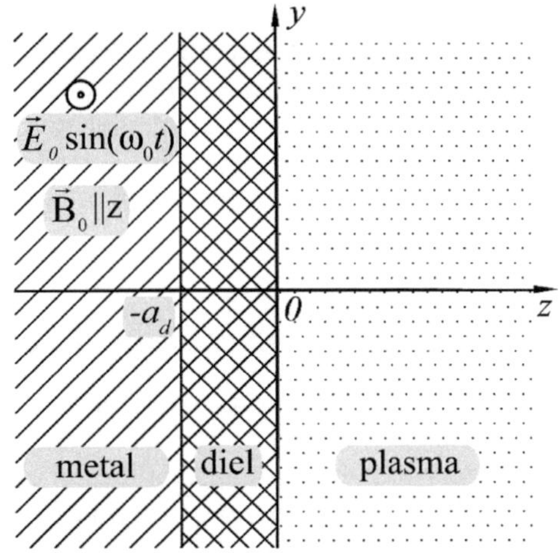


**Fig. 3.16** Schematic of the problem "parametric excitation of SCTM-modes"

transverse to the magnetic field $\vec{B}_0$. It is supposed here, that the amplitude of the external RF field is uniform. This assumption is valid under the condition that the RF wavelength is much longer than the wavelength of the electron SCTM-modes, and that the penetration depth of the external RF field into the plasma exceeds the penetration depth of the studied TM-modes.

The initial set of equations consists of the kinetic Vlasov-Boltzmann equations for the plasma particle distribution function that is supposed to be an equilibrium Maxwellian and the set of Maxwell equations for the fields of electron SCTM-modes. The dependence of the components of the electromagnetic field of the studied TM-modes on coordinates and time is assumed to be a sum of harmonics of the external RF field:

$$E, H = \sum_{n=-\infty}^{\infty} E^{(n)}, H^{(n)} \exp(ik_1 x - i\omega_n t), \tag{3.55}$$

where $\omega_n = \omega + n\omega_0$, and $E^{(n)}$ and $H^{(n)}$ are the Fourier harmonics of the corresponding components of the electron SCTM-mode field.

There is no dependence on the $y$ coordinate. The external constant magnetic field is supposed to be relatively weak, so that the inequality $\omega_e^2 < \Omega_e^2$ is satisfied. Then as it is shown in Sect. 3.1, in such waveguide systems electron SCTM-modes can propagate and their dispersion features have been studied in the previous sections. The problem of their propagation has been solved there in the approximation of slow electromagnetic eigenwaves under the condition of weak spatial dispersion of the plasma along the direction perpendicular to the plasma-dielectric interface.

The set of Maxwell equations for these modes with the electromagnetic field components $E_x$, $H_y$, $E_z$ (3.1) can be solved by the Fourier method. Since the fields of the studied TM-modes in the present problem depend on time and coordinates in a way shown in Formula (3.55), this circumstance allows one to obtain the following equation for the Fourier coefficients of the tangential component $E_x$ of the electric field of electron SCTM-modes, that includes not only the $n$-th but also the $(n + l)$-th Fourier harmonics of the fields of these modes:

$$\frac{2ck_1^2}{i\omega_n} H_y^{(n)}(0) + \left(k_1^2\psi_1^{(n)} + k_3^2\psi_2^{(n)}\right) E_1^{(n)}$$
$$= \sum_{\alpha} \sum_{\substack{s,m,l \\ l\neq 0}} \left\{ k_1^2 \left\{ \frac{s^2}{y_\alpha} + \frac{k_3^2(\omega_{n+m} + s\omega_\alpha)}{k_1^2(\omega_{n+m} - s\omega_\alpha)} \right\} \right.$$
$$\left. \frac{\Omega_\alpha^2 \exp(-y_\alpha) I_s(y_\alpha)}{\omega_n(\omega_{n+m} - s\omega_\alpha)} J_m(a_E) J_{m-l}(a_E) \right\} E_1^{(n+l)}. \tag{3.56}$$

In (3.56), the following notations are applied:

$$\psi_1^{(n)} \approx 1 - \sum_\alpha \sum_s \frac{s^2 \Omega_\alpha^2 \exp(-y_\alpha) I_s(y_\alpha)}{y_\alpha \omega_n(\omega_n - s\omega_\alpha)},$$
$$\psi_2^{(n)} \approx 1 - \sum_\alpha \sum_s \frac{\Omega_\alpha^2 (\omega_n + s\omega_\alpha) \exp(-y_\alpha) I_s(y_\alpha)}{\omega_n(\omega_n - s\omega_\alpha)^2},$$

$H_y^{(n)}(0)$ is the value of the tangential magnetic field component of the SCTM-mode on the plasma-dielectric interface. It should be underlined that the sum over the index $l$ does not include $l = 0$, and that the argument $a_E$ of the Bessel functions of the first kind has been determined in Sect. 2.3 to have the value $a_E = k_1 e_\alpha E_0/[(\omega_0^2 - \omega_\alpha^2)m_\alpha]$.

Now the boundary conditions for the tangential components of these TM-mode fields on the plasma-dielectric interface have to be formulated. The first of them is the commonly known linear boundary condition. It describes the continuity of the tangential electric field of these modes on the plasma-vacuum interface ($z = 0$ plane):

$$E_x^{(n)}(z = +0) = E_x^{(n)}(z = -0). \tag{3.57}$$

The second boundary condition can be obtained by integrating the Maxwell equations. Its physical meaning is the existence of the discontinuity of the tangential magnetic field $H_y^{(n)}(z)$ on the plasma-dielectric interface that is determined by the nonlinear surface electric current [38] that flows along the plasma boundary:

$$H_y^{(n)}(+0) - H_y^{(n)}(-0)$$
$$= \sqrt{\psi_1^{(n)}\psi_2^{(n)}} \sum_\alpha \sum_{\substack{s,m,l \\ l\neq 0}} \frac{s\omega_\alpha \Omega_\alpha^2 \exp(-y_\alpha) I_s(y_\alpha)}{ic|k_1|(\omega_{n+m} - s\omega_\alpha)^2} J_m(a_E) J_{m-l}(a_E) E_x^{(n+l)}(0) \tag{3.58}$$

Here the sum over the summation indices $s,\ m,\ l$ should be calculated separately over these three indices. So it can be considered to consist of three independent sums which are nested into each other. The summation procedure should be carried out starting from $-\infty$ until $+\infty$ for each sum. Again, the term related to $l = 0$ can be omitted in the sum over the $l$ index. The sum over the summation index $\alpha$ relates to all types of plasma particles (electrons and ions, respectively).

Carrying out the inverse Fourier transform, one can derive an infinite set of equations for the values of the tangential electric field component of the electron SCTM-modes calculated on the plasma boundary starting from (3.56) and applying the boundary conditions (3.57) and (3.58):

$$D_n(\omega, k_1)E_x^{(n)}(0) - \sum_{l,l\neq 0} F_{n,l}(\omega, k_1)E_x^{(n+l)}(0) = 0, \tag{3.59}$$

where

$$D_n(\omega, k_1) = 1 - \varepsilon_{\mathrm{d}} \coth(|k_1|a_{\mathrm{d}})(\psi_1^{(n)}\psi_2^{(n)})^{-1/2},$$

$$F_{n,l}(\omega, k_1) = \sum_{\alpha}\sum_{s,m} \frac{s\omega_\alpha \Omega_\alpha^2 \exp(-y_\alpha)I_s(y_\alpha)}{\omega_n(\omega_{n+m} - s\omega_\alpha)^2} J_m(a_E)J_{m-l}(a_E)\,.$$

Normally, the components of the factors $F_{n,l}(\omega, k_1)$ in (3.59) are small compared to the values of the corresponding term $D_n(\omega, k_1)$ because they include the product of two Bessel functions of the first kind $J_m(a_E)J_{m-l}(a_E)$. Only under specific conditions their values can be relatively large. Then they should be taken into account for an approximate theoretical analysis of the set of equations (3.59). This is the case if the following resonance condition is satisfied for the specific values of the summation indices $s$ and $m$:

$$\omega(k_1) \approx s \cdot |\omega_e| - m\omega_0. \tag{3.60}$$

Small resonant denominators appear in the expressions for the corresponding factors $F_{n,l}(\omega, k_1)$. As a result of satisfying this resonance condition the set of equations (3.59) obtains complex solutions, including those which describe unstable states of the studied plasma system.

If there is no external RF field, i.e. when $a_E = 0$, one can derive the expression for the dispersion relation $D(\omega, k_1) = 0$ from the system of equations (3.59). This relation allows one to calculate the eigenfrequency $\omega_{\mathrm{eig}}$ of the electron SCTM-modes propagating in the considered plasma structure. If an external RF field is applied, $a_E \neq 0$, under the validity of the resonance relation (3.60), this set of equations describes the parametric instability of electron SCTM-modes.

In the case of small amplitude plasma electron waves caused by an external RF field, namely when this amplitude is smaller than the electron Larmor radius, the argument $a_E$ of the first kind Bessel functions in expression (3.59) is small ($a_E < 1$). Under this condition one can find the expression for the growth rate of the parametric instability of electron SCTM-modes analytically. It is assumed that the angular frequency $\omega$ of these modes in the case of utilization of the pumping electric field $\vec{E}_0 \sin(\omega_0 t)$ differs slightly from the eigenfrequency $\omega_{\mathrm{eig}}$ of these modes propagating in the studied waveguide system under the condition of absence of an external RF field. Then one can search for the solution of the dispersion relation (3.59) in the form $\omega = \omega_{\text{eig}} + \gamma$ and assume that the inequality $|\gamma| \ll \omega_{\text{eig}}$ is valid. Since $a_E$ is proportional to the amplitude of the pumping electric field ($a_E \propto E_0$), under the approximation of a weak amplitude of the pumping RF field, one can derive from (3.59) the following dispersion relation:

$$D_0 - \left\{ \frac{\Omega_e^2 \exp(-y_e) I_s(y_e)}{(\gamma + \Delta_{Te})^2} \right\}^2 \frac{s\omega_e \omega_{\text{eig}}}{\omega_{\text{eig}}^2 - \omega_0^2} \cdot \frac{a_E^2}{2} = 0, \quad (3.61)$$

where $\omega_{\text{eig}} = s \cdot |\omega_e| + \Delta_s$.

The term $\Delta_{Te} = \Delta_{Te}(s, \upsilon_{Te})$ in the denominator determines the shift of the eigenfrequency of electron SCTM-modes relative to the corresponding harmonic of the electron cyclotron frequency. In the case of arbitrary values of the studied waveguide parameters and wavelength of the SCTM-modes, the value of $\Delta_s$ can be found from numerical analysis. But in the mutually opposite limiting cases, namely for thick ($|k_1|a_{\text{d}} \gg 1$) and thin ($|k_1|a_{\text{d}} \ll 1$) dielectric layers, and also for ranges of short ($y_e \gg 1$) and long ($y_e \ll 1$) wavelengths of the electron SCTM-modes one can find simple analytical expressions for the parameter $\Delta_{Te}$. Doing that one can use its relation with the parameter $h_e$, that was calculated in the previous Sects. 3.1 and 3.2 where the dispersion properties of these modes have been analysed in detail: $\Delta_{Te} = s|\omega_e| \cdot h_e$.

Thus, if the inequality $|k_1|a_{\text{d}} \gg 1$ is valid, which can be interpreted as the limiting case of a thick dielectric layer separating the plasma region from the metal wall of the waveguide, one can apply Formulas (3.5) and (3.6), which are found in Sect. 3.1 for the ranges of long and short wavelengths, respectively. Under the resonance condition $\omega = \omega_{\text{eig}} + \gamma$, one can derive from (3.61) the following algebraic equation of fifth order in respect to the correction $\gamma$ [38]:

$$\gamma(\gamma + \Delta_s)^4 = \left\{ \Omega_e^2 \exp(-y_e) I_s(y_e) \right\}^2 \frac{s\omega_e \omega_{\text{eig}}}{\omega_{\text{eig}}^2 - \omega_0^2} \cdot \frac{a_E^2}{2} \left( \frac{\partial D_0}{\partial k_1} \right)^{-1} \upsilon_{\text{gr}}, \quad (3.62)$$

where $\upsilon_{\text{gr}}$ is the group velocity of these modes. Solving of (3.62) allows one to derive analytical expressions for the growth rates of the parametric instability of electron SCTM-modes. One can find a simple analytical expression for this growth rate $\text{Im}\gamma$ in the case when the inequality $|\gamma| \gg |\Delta_{Te}|$ is valid. Then in the case of a weak plasma spatial dispersion along the direction of propagation of these modes ($y_\alpha \ll 1$), the growth rate $\text{Im}\gamma$ is determined by the following formula:

$$\text{Im}\gamma \approx |\Delta_{Te}| \left\{ \frac{y_e^2 \varepsilon_{\text{d}}^{-1}}{I_s(y_e) 2(s^2 - 1)^2} \right\}^{1/5} a_E^{2/5}. \quad (3.63)$$

In this case, the inequality $y_e^2 a_E^2 \gg 2s^4 I_s(y_e)$ is assumed to be valid. In the range of moderate values of wavelengths, $y_e \ll 1 \ll y_i$, one can apply Formula (3.63) for deriving analytical expression of the growth rate $\text{Im}\gamma$.

Under the condition of a strong plasma spatial dispersion along the plasma interface ($y_e \gg 1$), the expression for the parametric instability growth rate $\text{Im}\gamma$ is as follows:

$$\text{Im}\gamma \approx |\Delta_{Te}| \left[ \frac{y_e \omega_{\text{eig}}}{\Delta_{Te} \varepsilon_{\text{d}}} \right]^{1/5} a_E^{2/5}. \tag{3.64}$$

The approximation $|\gamma| \gg |\Delta_{Te}|$, that was applied for obtaining the expressions (3.63) and (3.64), can be satisfied in the most easy way just for the electron SCTM-modes with short wavelengths, for which the following inequality is valid: $y_e \gg \Delta_{Te} s^2 a_E^{-2} \omega_{\text{eig}}^{-1}$.

As it is shown in Sect. 3.3, a decrease of the thickness of the protective dielectric coating located on the inner wall of the metal waveguide, where the studied TM-modes can propagate, leads to deterioration of the propagation conditions of these eigenmodes. Therefore, one can foresee a negative influence of a decrease of the thickness of the dielectric coating on the value of the frequency correction $\gamma$ for all the limiting cases investigated above.

Thus in the case of a thin dielectric layer ($a_{\text{d}} k_1 \ll 1$), one can apply Formulas (3.12) and (3.13) for determination of an analytical expression for the frequency shift $h_e$ in the ranges of the SCTM-mode wavelengths mentioned above. Since the asymptotic expression for the hyperbolic cotangent differs in this case compared with the previous case of a thick dielectric layer, namely, $\coth(k_1 a_{\text{d}} \ll 1) \sim (a_{\text{d}} k_1)^{-1}$, an initial stage of parametric instability of the electron SCTM-modes is described by another equation than (3.62). This is given by:

$$\gamma(\gamma + \Delta_{Te})^4 = \frac{\Omega_e^2 I_s(y_e) y_e \omega_{\text{eig}} a_E^2 \omega_e^2 \Delta_{Te}^2 \exp(-y_e)}{\psi_2^{(0)}(a_E = 0) \cdot cth(k_1 a_{\text{d}})(\omega_{\text{eig}}^2 - \omega_0^2)}. \tag{3.65}$$

In the approximation of a thin dielectric layer ($a_{\text{d}} k_1 \ll 1$), the expressions for $\Delta_{Te}(y_e)$ differ from those obtained in the case of a thick dielectric layer. Therefore, the initial stage of the parametric instability of electron SCTM-modes is characterized by other expressions for growth rates, whose values are smaller than those found for the previous limiting case of a thick dielectric layer. Thus in the range of long wavelengths, one can find the following analytical expression for the growth rate of parametric instability of these TM-modes:

$$\text{Im}\gamma \approx |\Delta_{Te}| \left\{ \frac{y_e^2 k_{\text{d}}}{I_s(y_e) 2 (s^2 - 1)^2} \right\}^{1/5} a_E^{2/5}, \tag{3.66}$$

where $k_{\text{d}} = |k_1| a_{\text{d}} / \varepsilon_{\text{d}} \ll 1$.

Correspondingly, in the range of short wavelengths, the parametric instability of these TM-modes is characterized by the following growth rates:

$$\mathrm{Im}\gamma \approx |\Delta_{Te}| \, [y_e \omega_{\mathrm{eig}} k_{\mathrm{d}} / \Delta_{Te}]^{1/5} a_E^{2/5}. \tag{3.67}$$

Thus, it is proved here that a decrease of the thickness of the dielectric coating leads to decreasing values of the growth rates of the parametric instability of electron SCTM-modes as it was predicted above.

Analysis of Formulas (3.66) and (3.67) allows one to make the conclusion that, first, parametric excitation of these TM-modes happens faster under the condition of a strong spatial dispersion along the plasma interface ($y_e \gg 1$) as compared with the case of a weak spatial dispersion ($y_e \ll 1$). Second, comparing the values of the parametric instability growth rates of these TM-modes with values of growth rates of the resonant beam-plasma instability of these modes, whose expressions have been obtained in the previous Sect. 3.4, one can make the conclusion that the efficiency of beam excitation is lower than the efficiency of parametric excitation under the condition of sufficiently weak density of the utilized charged particle beam: $a_E^{2/5} > (n_b/n_p)^{1/3}$. Finally, the dependence of the parametric growth rate of electron SCTM-modes $\mathrm{Im}\gamma$ on the amplitude $a_E$ of the pumping RF field differs from that found in the case of longitudinal quasi-potential bulk cyclotron waves [36, 37], where the growth rate is $\mathrm{Im}\gamma \propto a_E^{2/3}$.

To complement the analytical investigations of the initial stage of parametric excitation of electron SCTM-modes one can solve (3.59) numerically. Affected by an external RF field the studied modes propagate as a wave packet, so one can take into the consideration one definite harmonic and its two nearest satellites among various harmonics of these TM-modes. Then instead of an infinite set of equations one can examine a set of three equations for the tangential electric field of the main harmonic SCTM-mode $E_x^{(0)}$ and its nearest satellites $E_x^{(\pm 1)}$. This set of equations is uniform and thus has a solution in the case, when the determinant constructed from the coefficients of the nearby harmonics of the tangential electric field is equal to zero. This determinant, that has to be solved numerically in order to study this parametric instability, is given by:

$$\begin{vmatrix} D_{-1} & F_{-1;+1} & F_{-1;+2} \\ F_{0;-1} & D_0 & F_{0;+1} \\ F_{+1;-2} & F_{+1;-1} & D_{+1} \end{vmatrix} = 0. \tag{3.68}$$

The results of the numerical analysis of equation (3.68) are presented in Figs. 3.17, 3.18 and 3.19. They demonstrate good agreement with the corresponding analytical results and allow one to obtain some additional information on the initial stage of the parametric instability. Carrying out the numerical analysis, one separates the expression for the argument of the Bessel functions of the first kind into two factors. The first of them describes only the spatial dispersion of the plasma along its interface,

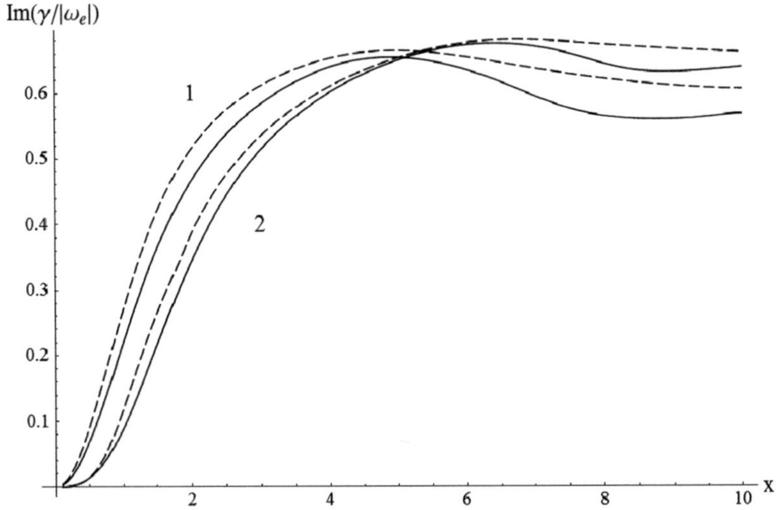


**Fig. 3.17** Growth rate of electron SCTM-modes versus the product $x = k_1\rho_e$ for $b_E = 0.5$ and $\omega_0 = |\omega_e|/2$. Solid and dashed curves relate to $Z = \Omega_e^2/\omega_e^2 = 1.5$ and 10, respectively. The numbers 1 and 2 denote the number of the cyclotron harmonic $s$=1 and 3, respectively

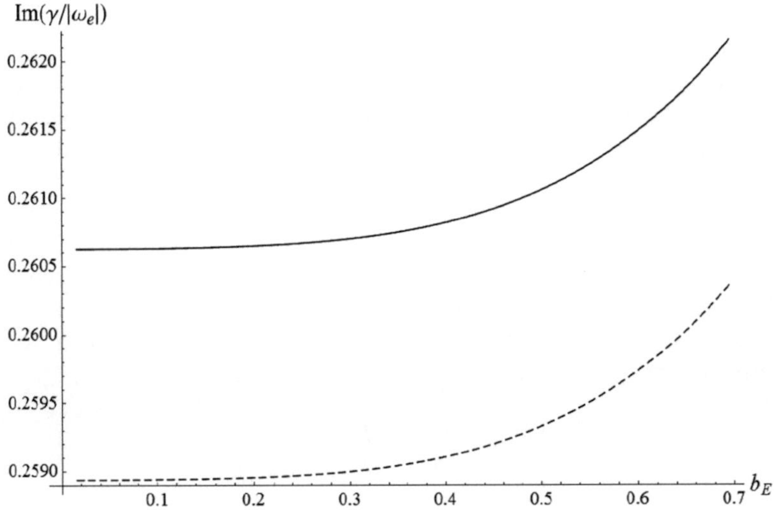


**Fig. 3.18** Growth rate of the electron SCTM-mode inside the first frequency band $1 < \omega/|\omega_e| < 2$ versus the dimensionless amplitude of the pumping electric field $b_E$ for $\omega_0 = |\omega_e|/2$ and $k_1\rho_e = 1$. Solid and dashed curves relate to $Z = \Omega_e^2/\omega_e^2 = 50$ and 100, respectively

namely, $y_e$, and the second one describes exclusively the amplitude of oscillations of the plasma electrons affected by an external RF field $\vec{E}_0(t)$:

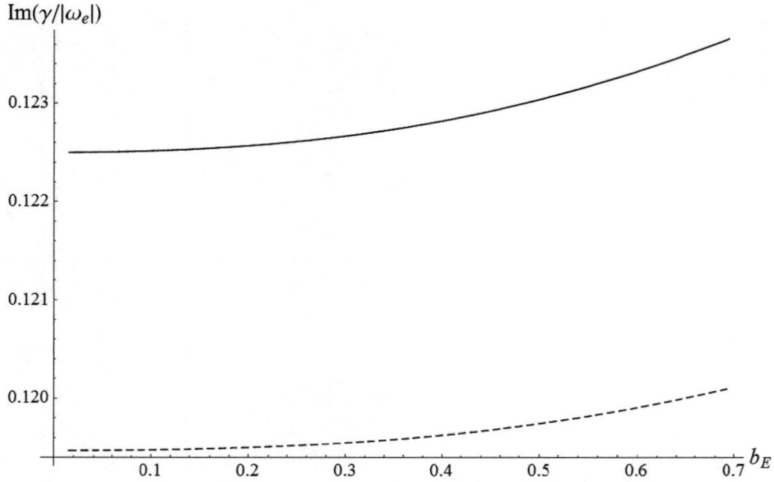


**Fig. 3.19** Growth rate of the electron SCTM-mode inside the second frequency band $2 < \omega/|\omega_e| < 3$ versus the dimensionless amplitude of the pumping electric field $b_E$ for $\omega_0 = |\omega_e|/2$ and $k_1\rho_e = 1$. Solid and dashed curves relate to $Z = \Omega_e^2/\omega_e^2 = 50$ and 100, respectively

$$a_E^2 = \frac{e^2 k_1^2 E_0^2}{m_e^2 \left(\omega_0^2 - \omega_e^2\right)^2} = 2 y_e b_E^2. \tag{3.69}$$

Carrying out the numerical analysis, one operates here with dimensionless parameters, namely, dimensionless growth rate $\mathrm{Im}(\gamma/|\omega_e|)$, dimensionless wave vector, which is denoted as $x = k_1\rho_e$, and the plasma density is described by the dimensionless parameter $Z = \Omega_e^2/\omega_e^2$.

The dependence of the parametric instability growth rate $\mathrm{Im}(\gamma/|\omega_e|)$ on the wavenumber and the plasma density is presented in Fig. 3.17. In the tested cases $\mathrm{Im}(\gamma/|\omega_e|)$ increases with decreasing plasma density (parameter Z) and increasing wavenumber (product $x = k_1\rho_e$). However, after $x \approx 5$ this dependence becomes of non-monotonous character. It is found that with increasing electron cyclotron harmonic number the curves $\mathrm{Im}(\gamma/|\omega_e|)$ are shifted as a whole to the side of higher values of the parameter $x$, which describes the plasma dispersion along the propagation direction of these TM-modes.

The dependences of the electron SCTM-mode growth rates on the dimensionless amplitude $b_E$ of an external RF field are shown in Figs. 3.18 and 3.19 for the first and second electron cyclotron harmonic, respectively. As it is mentioned above, in order to identify the contribution only of the amplitude of the applied RF field it is convenient to take just the parameter $b_E = a_E/\sqrt{2y_e}$.

Regardless of the choice of the waveguide parameters the value of $\mathrm{Im}(\gamma/|\omega_e|)$ increases with increasing parameter $b_E$. One can also see in both Figs. 3.18 and 3.19 that changing the plasma density does not essentially influence the shape of the curves. Unlike this, changing the electron cyclotron harmonic number $s$ from 1 to 2 in

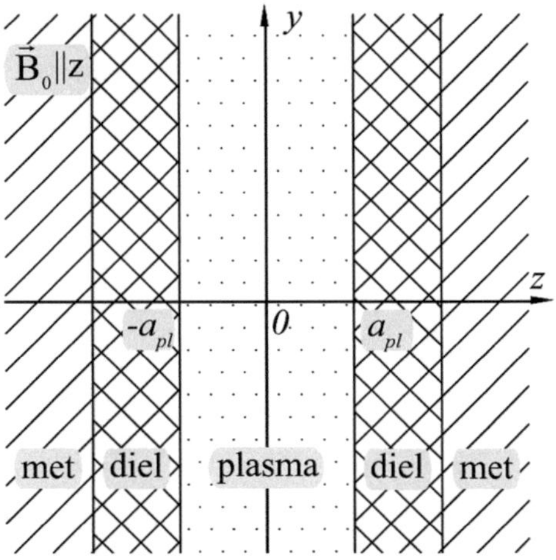


**Fig. 3.20** Schematic of the problem "influence of finite plasma layer thickness on SCTM-mode parametric instability"

general leads to a decrease of the absolute values of the parametric instability growth rates more than by a factor of two and to a reduction of the growth rate $\mathrm{Im}(\gamma/|\omega_e|)$ with increasing dimensionless wave vector.

Since it has been found that the thickness of the plasma layer essentially changes the dispersion of the studied TM-modes, it is interesting now to investigate the influence of a finite size of the plasma layer thickness on the initial stage of the electron SCTM-mode parametric instability. A planar metal waveguide with the following geometrical parameters is considered (see Fig. 3.20). The space $-a_{\mathrm{pl}} \leq z \leq a_{\mathrm{pl}}$ is filled with a gyrotropic plasma layer that is limited by a dielectric medium with a dielectric permittivity that is approximately equal to unit (for simplification of our considerations). The thickness of the dielectric layer is assumed to be sufficiently large to apply a model of semibounded space for its description. Then, with the aid of the method applied for the limiting case of semibounded plasmas, one can derive the following infinite set of equations for the tangential component of the electric field at the plasma layer interface:

$$D_n E_x^{(n)}(x = a_{\mathrm{pl}}) - \sum_{\alpha} \sum_{\substack{s,m,l \\ l \neq 0}} F_{n,l} E_x^{(n+l)}(x = a_{\mathrm{pl}}) \approx 0, \tag{3.70}$$

where

$$D_n = (\psi_1^{(n)} \psi_2^{(n)})^{-1/2} - \mathrm{th}\left( a_{\mathrm{pl}} k_1 \sqrt{\psi_1^{(n)} / \psi_2^{(n)}} / 2 \right),$$

$$F_{n,l} = \sum_{\alpha} \sum_{\substack{s,m,l \\ l \neq 0}} \frac{\Omega_\alpha^2 I_s(y_\alpha) s \omega_\alpha J_m(a_E) J_{m-l}(a_E)}{(\omega_{n+m} - s\omega_\alpha)\omega_n \exp(y_\alpha)} \quad .$$

The sum over the index $\alpha$ should be calculated for all species of plasma particles, and the sums over the indices $l, m, s$ have to be independently calculated in the limits from $-\infty$ to $+\infty$ for all natural numbers (for the index $m$ one should take into account as well the magnitudes of the terms calculated for its zero value).

This set of equations can be used both in the cases of ion and electron SCTM-modes. When considering the case of electron TM-modes, one can neglect ion contributions in the sum presented in (3.70) because of their smallness. Analytically the problem can be solved in the limiting cases of thick ($a_{\rm pl}k_1 \gg 1$) and thin ($a_{\rm pl}k_1 \ll 1$) plasma layers, using asymptotic expansions in series of hyperbolic functions over their arguments. It can be predicted before starting the analysis, that the set of equations for the tangential electric field of the TM-modes in the limiting case of a thick plasma layer is equivalent to the set of equations derived previously for the model of semibounded plasmas. That is why this case is not studied in details. However, in contrast to this, the limiting case of a thin plasma layer is analyzed in details here. Doing that one assumes that even the wavelength of the electron SCTM-modes that belongs to the short wavelength range is larger than the thickness of the plasma layer, thus the inequality $a_{\rm pl} \ll 2\pi\, k_1^{-1}$ is satisfied for the considerations presented in the next part of this section.

Here, one considers the following resonance condition: $\omega = \omega_{\rm eig} + \gamma$, where $\omega_{\rm eig} = s|\omega_e| + \Delta_{Te}(s, v_{Te})$. It is assumed as well, that the following inequality $\omega_{\rm eig} \gg |\gamma| \gg |\Delta_{Te}|$ is valid. Under this resonance condition, finding the solutions to the set of equations (3.70) is reduced to analyzing the determinant constructed from the coefficients located nearby the harmonics of the tangential electric field $E_x^{(n+l)}(a_{\rm pl})$ of these TM-modes calculated on the plasma interface. Taking into account only the main harmonic of this field $E_x^{(n)}(a_{\rm pl})$ and its two nearest sub-harmonics $E_x^{(n\pm1)}(a_{\rm pl})$ one can derive the following equation for the frequency correction $\gamma$ of the TM-modes:

$$(D_0 D_1 - F_{0,1} F_{1,-1}) D_{-1} - (D_1 F_{0,-1} + F_{0,1} F_{1,-2}) F_{-1,1} \approx 0. \tag{3.71}$$

In the limiting case of very long wavelengths ($y_e < y_i \ll 1$) the following inequality is valid:

$$\left|a_{\rm pl} k_1\right| \ll a_E^{4/5} \left(\Omega_e/\omega_{\rm eig}\right)^{6/5}, \tag{3.72}$$

and the growth rate of the parametric instability of electron SCTM-modes is described by the following formula:

$$\mathrm{Im}\gamma \approx \omega_{\rm eig} \left(\frac{\Omega_e}{\omega_{\rm eig}}\right)^{3/5} \frac{a_E^{2/5}}{\sqrt{a_{\rm pl} k_1}} \left(\frac{y_e^s}{2^s s!}\right)^{1/2}. \tag{3.73}$$

In the case when the spatial dispersion for plasma electrons is weak but at the same time it is strong for the plasma ions, so that the inequality $y_e \ll 1 \ll y_i$ is valid, the asymptotes of the functions $\psi_1^{(n)}$ and $\psi_2^{(n)}$ have another form. Then the coefficients in (3.71), that is applied for finding the values of the frequency shift $\gamma$, transform as well. Under the condition of satisfaction of the inequality:

$$\left|a_{\text{pl}}k_1\right| \ll a_E^{4/5}\left(\frac{4\Omega_e^3}{\omega_{\text{eig}}^3}\right)^{2/5}\frac{y_e^2}{s^3 I_s(y_e)}, \tag{3.74}$$

the growth rate of this parametric instability is equal to:

$$\text{Im}\gamma \approx \omega_{\text{eig}}\left(\frac{4\Omega_e^3}{\omega_{\text{eig}}^3}\right)^{1/5}\frac{a_E^{2/5}}{\sqrt{a_{\text{pl}}k_1}}\left(\frac{y_e^s}{2^s s!}\right)^{1/2}. \tag{3.75}$$

In the range of short wavelengths of electron SCTM-modes, if the inequality

$$\left|a_{\text{pl}}k_1\right| \ll a_E^{4/5}\frac{y_e^{5/2}}{s^2}\frac{\omega_e^2}{\Omega_e^2} \tag{3.76}$$

is valid, the growth rate of the parametric instability of these TM-modes reaches the maximum magnitude as it was in the case of the previous model (semibounded plasma):

$$\text{Im}\gamma \approx \frac{a_E^{2/5}\Omega_e}{\sqrt{a_{\text{pl}}k_1 y_e^{1/4}}(4\pi)^{1/5}}. \tag{3.77}$$

Comparison of the results obtained for the model of a narrow plasma layer with the results obtained for the model of a semibounded plasma shows that the value of the growth rate of the parametric instability of SCTM-modes is larger for narrow layers. This can be explained by the circumstance that the large geometrical factor $(a_{\text{pl}}k_1)^{-1/2} \gg 1$ appears in Formulas (3.73), (3.75) and (3.77), but not in the formulas for the growth rates of electron SCTM-modes propagating along the interface of a semibounded plasma.

It should be pointed out as well, that there exists a critical value of the relative velocity of plasma electron oscillations due to the application of an external RF pumping field that must be exceeded to realize the development of parametric instabilities of SCTM-modes. The threshold value $a_{\text{th}}$ can be estimated from the condition that the growth rate of electron SCTM-mode parametric instabilities is equal to the damping rate of these modes determined mainly by collisions between the plasma particles. In the case of a weak plasma spatial dispersion along its interface ($y_e \ll 1$), one can find the following expression for the threshold value of the amplitude of the RF field $a_E$:

$$a_{\text{th}}^{2/5} \approx \frac{\nu_e}{\omega}\left(\frac{\omega_e}{\Omega_e}\right)^{3/5}\sqrt{\frac{a_{\text{pl}}s!}{2^S\rho_e}}(k_1\rho_e)^{0.1-S}. \tag{3.78}$$

In the case of strong spatial dispersion ($y_e \gg 1$) one can find:

$$a_{\rm th}^{2/5} \approx \frac{\nu_e}{\Omega_e}\sqrt{\frac{a_{\rm pl}}{\rho_e}}(k_1\rho_e)^{3/5}. \tag{3.79}$$

These threshold values of $a_E$ can be easily reached in the case of weakly collisional plasmas ($\nu_e \ll \Omega_e$). With the development of the parametric instability, the value of the plasma particle collision frequency caused by the turbulent oscillations of electromagnetic fields begins to increase. Finally, when the value of the effective collision frequency $\nu_e$ of the plasma particles reaches the value of the parametric instability growth rate, the instability passes to the stage of saturation. After that, the instability develops nonlinearly. However, the present theory does not describe this process.

Summarizing the results presented in this section it should be pointed out that electron SCTM-modes can be parametrically excited in the most effective way at frequencies of an external RF fields, which are close to eigenfrequencies of these modes. Analytical expressions for the electron SCTM-mode growth rates show that they differ from expressions obtained for the bulk electron cyclotron waves [37]. The amplitude of the external RF field, the wavelength of the studied modes and their electron cyclotron harmonic number mainly influence the initial stage of this parametric instability. Decreasing the mode's wavelength and increasing the amplitude of the external RF field lead to an increase of their growth rates. The growth rates of parametric instabilities can also be increased by decreasing the plasma layer thickness.

The obtained results can be useful, first, for the development of plasma technologies based on utilization of surface electron cyclotron waves. It is well-known that application of surface waves has many advantages compared with the use of bulk waves for sustaining gas discharges with large operating surface and for uniform plasma production [39]. Second, they can be useful for diagnostics of the periphery of fusion plasmas [40], and for searching a possibility to decrease plasma periphery heating in thermonuclear fusion plasma devices. Finally, we would like to point out a new field of application of the obtained theoretical knowledge in the branch of classical electrodynamics of bounded plasma, namely, in plasmonics (see [41] and references therein). This monograph studies the collective motion of conductivity electrons in metal nano-structures excited by electromagnetic waves, whose wavelength belongs to the range of visible light.

## 3.6 Excitation of Surface Electron Cyclotron TM-Modes by Non-monochromatic External RF Fields

Often, plasmas are under the influence of electric fields radiated by several RF generators which operate at different frequencies or the frequency spectrum of their radiation is strongly non-monochromatic. Therefore, the problem of the influence of non-monochromatic external RF fields on confined magnetized plasmas is of great importance. Experimentally, the influence of electromagnetic waves from generators which operate at different frequencies was studied in e.g. [42, 43]. It was found that the scenario of parametric instability in this case essentially differs from that in the case of application of a monochromatic pumping RF field. Theoretically, instability of an unbounded plasma under the influence of two RF fields with different frequencies was studied in [44]. This research has allowed one to establish clearly that in the presence of significant non-monochromatic external RF fields, the development of parametric plasma instabilities occurs in a different way as compared with the case of applying a monochromatic RF field. Moreover, the growth rates of this instability become strongly non-monotonous functions depending on various plasma parameters, where this instability could be observed. Unfortunately, the case of parametric excitation of surface waves by a two-frequency pumping force was studied in details only for waves propagating in an incompressible liquid, but not in plasmas, see e.g. [45] and references therein.

Interesting experimental results which confirm the mentioned conclusion on parametric instabilities excited by two-frequency pumping RF fields were obtained during the experiments on controlled thermonuclear fusion performed at the torsatron "Uragan-3" [46]. The parameters of the device were as follows: magnetic field $B_0 \leq 1$ T, major radius of helical winding $R_0 = 1$ m, minor radius of helical winding $r = 0.27$ m, and average plasma radius $a \approx 0.135$ m. The utilized RF power sources were two independent generators with the different frequencies $f_1 = 2.3$ MHz and $f_2 = 5.3$ MHz, the output powers $P_1 \approx P_2 \approx 500$ kW, and the pulse durations $\tau_1 \approx \tau_2 \approx 50$ ms. These experiments showed that use of two RF generators with different frequencies results in obtaining significantly different plasma parameters compared with the case of a monochromatic pumping RF field. The theory of parametric plasma instability in the case of two pumping waves of small amplitude under the condition that their frequencies are close to the electron plasma frequency and their frequency difference is close to the frequency of ion sound waves was developed in [47]. The authors showed that under parametric excitation of ion-acoustic oscillations, the presence of a second pumping RF field can lead to either increasing or decreasing growth rates of parametric instabilities of these waves. Application of this theory allows one to explain the experimental results presented in [42, 43].

In paper [47], the theory of parametric instability of bulk ion cyclotron waves, which propagate in an unbounded plasma, was generalized for the case of finite oscillation amplitude of the charged plasma particles influenced by two external RF fields with different operating frequencies. The ratio of the operating frequencies was chosen to be:

$$\omega_{01}/\omega_{02} = p_1/p_2, \tag{3.80}$$

where $p_1$ and $p_2$ are natural numbers. Thus, in this case, one can get the following expression for the operating frequencies:

$$n_1\omega_{01} + n_2\omega_{02} = N_0\omega_0, \tag{3.81}$$

where $n_1 p_1 + n_2 p_2 = N_0$, and $(\omega_{01} - \omega_{02})/(p_1 - p_2) = \omega_0$.

Therefore, under the conditions of validity of the relations (3.80) and (3.81) one can reduce the problem of excitation of electromagnetic waves in plasmas which are influenced by two different RF fields to the problem of parametric instability of plasma waves excited by a single pumping RF field with an effective operating angular frequency $\omega_0$.

One of the main results of paper [47] which was devoted to the case of excitation of bulk ion cyclotron waves is the following conclusion. If the operating frequency of one of the pumping waves, for example, $\omega_{01}$, is close to the ion cyclotron frequency, the dependence of the growth rate of the parametric instability of these waves on amplitude and operating frequency of the second pumping RF field becomes non-monotonous. Thus by means of changing just these parameters of the complex non-monochromatic pumping RF field one can achieve either the regime of stabilization or a regime of enhancing the parametric instability.

However, as mentioned in the previous section, the theory of parametric instabilities of surface cyclotron waves is developed weakly compared with the case of bulk waves. Therefore, this section is devoted to the theory of parametric excitation of surface electron cyclotron TM-modes in the case of a non- monochromatic pumping RF field. Doing that, the effect of the non-monochromatic external RF field is modelled by a superposition of two RF fields with different amplitudes and operating frequencies.

In the following this problem is analysed applying the model of semibounded plasmas with uniform density (see Fig. 3.21). It is assumed that the plasma occupies the half-space $z \geq 0$, which is separated from the metal wall of the waveguide by a dielectric layer with dielectric permittivity $\varepsilon_d$ and thickness $a_d$. In the region of the considered waveguide space $z < -a_d$, the metal wall of the waveguide is located, its conductivity is assumed to be sufficiently large to allow the assumption that the tangential component of the SCTM-mode electric field is equal to zero at the surface of the metal wall. An external static magnetic field is oriented along the normal to the plasma interface $\vec{B}_0 \| \vec{z}$, and its magnitude is sufficiently weak so that the following inequality between electron cyclotron frequency and electron plasma frequency is valid: $\omega_e^2 < \Omega_e^2$. We assume as well that the electric field of both external pumping RF waves has not any component along the static magnetic field $\vec{B}_0$ and that the spatial distribution of their fields is quite uniform, which is true under the conditions that the wavelengths of the two pumping waves are much longer than the wavelength and penetration depth of the studied TM-modes.

Theoretical study of the parametric instability of SCTM-modes is carried out here under the condition that the electric field of the non-monochromatic RF pumping waves is simulated by superposition of two RF fields with different amplitudes and operating frequencies:

$$\vec{E}_0(t) = \vec{E}_{01}\sin(\omega_{01}t + \beta_1) + \vec{E}_{02}\sin(\omega_{02}t), \tag{3.82}$$

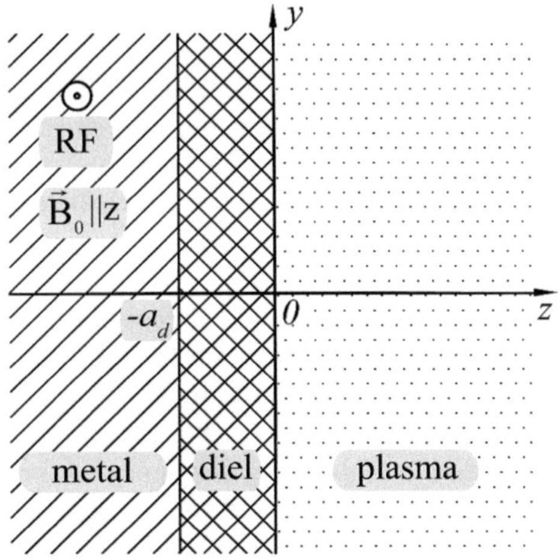


**Fig. 3.21** Schematic of the problem "Excitation of SCTM-modes by non-monochromatic external RF fields"

where $E_{01,02}$ are the amplitudes of the electric components of the RF field, $\omega_{01,02}$ are the operating frequencies and $\beta_1$ is the starting value of the phase of the first component of this non-monochromatic pumping RF field.

The set of equations, which describes the process of SCTM-mode parametric excitation, consists of the kinetic Vlasov-Boltzmann equation for the disturbed distribution function of the plasma particles and the set of Maxwell equations for the fields of the studied modes. The dependences of the SCTM-mode fields on space coordinates and time are assumed to be in the form of a sum over Fourier harmonics of the considered non-monochromatic external RF field:

$$E, H = \sum_{m_1, m_2=-\infty}^{+\infty} E(m_1, m_2), H(m_1, m_2) \exp(ik_1 x - it\omega(m_1, m_2)), \quad (3.83)$$

where $\omega(m_1, m_2) = \omega + m_1\omega_{01} + m_2\omega_{02}$.

There is no dependence of the TM-mode fields on the coordinate $y$. The problem is considered in the approximations of weak spatial dispersion of the plasma along the direction perpendicular to the plasma interface, and of a relatively slow phase velocity of SCTM-mode propagation compared with the velocity of electromagnetic waves in vacuum. The dispersion properties of these modes have been studied in the Sects. 2.1 and 2.2.

The kinetic Vlasov-Boltzmann equation for the first correction to the distribution function of the plasma particles in this case has the following form:

$$\frac{\partial f_{1\alpha}}{\partial t} + \vec{v}\frac{\partial f_{1\alpha}}{\partial \vec{r}} + \frac{e_\alpha}{m_\alpha}\left\{\vec{E}_{01}\sin(\omega_{01}t + \beta_1) + \vec{E}_{02}\sin(\omega_{02}t) + \frac{1}{c}[\vec{v}, \vec{B}_0]\right\}\frac{\partial f_{1\alpha}}{\partial \vec{v}}$$

$$= -\frac{e_\alpha}{m_\alpha}\left[\vec{E}(t) + \frac{1}{c}[\vec{v}, \vec{H}]\right]\frac{\partial f_{0\alpha}}{\partial \vec{v}}. \quad (3.84)$$

The distribution function of the plasma particles under the equilibrium stage $f_{0\alpha}$ is determined by the following Maxwellian distribution function:

$$f_{0\alpha}(\vec{v}, t) = f_M\left\{\vec{v} + \frac{e_\alpha}{m_\alpha}\left[\frac{\vec{E}_{01}}{\omega_{01}}\cos(\omega_{01}t + \beta_1) + \frac{\vec{E}_{02}}{\omega_{02}}\cos(\omega_{02}t)\right]\right\}. \quad (3.85)$$

The method of solving the kinetic equation (3.84) is described in details in Sect. 2.4. One can apply expression (2.101) for the electrical conductivity tensor of the gyrotropic plasma that is influenced by the external non-monochromatic RF field (see Formula (3.82)) without any additional explanation. This allows one to write down the expressions for the Fourier coefficients of the electric current density in the plasma in the approximation of weak spatial dispersion of the plasma in the direction perpendicular to the plasma interface. They deliver the set of equations that describes the studied TM-mode:

$$\begin{aligned} j_1(q_1, q_2) = \sum_\alpha \frac{i\Omega_\alpha^2}{4\pi} \sum_{s,m1,m2,p1,p2} \frac{I_s(y_\alpha)\exp(-y_\alpha)}{\exp(-i\beta_1 p_1)} G(m_1, m_2, p_1, p_2) \\ \times \left[\frac{s^2 E_1(p_1+q_1, p_2+q_2)}{[\omega(m_1+p_1, m_2+p_2) - s\omega_\alpha]y_\alpha}\right. \\ \left. + \frac{s\omega_\alpha k_3 E_3(p_1+q_1, p_2+q_2)/k_1}{[\omega(m_1+p_1, m_2+p_2) - s\omega_\alpha]^2}\right], \end{aligned} \quad (3.86)$$

$$\begin{aligned} j_3(q_1, q_2) = \sum_\alpha \frac{i\Omega_\alpha^2}{4\pi} \sum_{s,m1,m2,p1,p2} \frac{I_s(y_\alpha)\exp(-y_\alpha)}{\exp(-i\beta_1 p_1)} G(m_1, m_2, p_1, p_2) \\ \times \left[\frac{s\omega_\alpha k_3 E_1(p_1+q_1, p_2+q_2)/k_1}{[\omega(m_1+p_1, m_2+p_2) - s\omega_\alpha]^2}\right. \\ \left. + \frac{E_3(p_1+q_1, p_2+q_2)}{\omega(m_1+p_1, m_2+p_2) - s\omega_\alpha}\right]. \end{aligned} \quad (3.87)$$

Here the following notations are applied:

$$G(m_1, m_2, p_1, p_2) = J_{m_1}(a_{E1})J_{m_1-p_1}(a_{E1})J_{m_2}(a_{E2})J_{m_2-p_2}(a_{E2}),$$

where the $a_{Ej} = \left|e_\alpha k_1 E_{0j}/[(\omega_{0j}^2 - \omega_\alpha^2)m_\alpha]\right|$ are the arguments of the first kind Bessel functions $J_n(a_{Ej})$.

Using (3.83) one can derive the expression for the Fourier coefficient of the tangential component of the electric field of the electron SCTM-modes from the Maxwell equations:

$$E_1(k_3, q_1, q_2) = \frac{ik_1^2 H_y(q_1, q_2)|_{z=+0}}{\pi \cdot k \cdot \Delta(k_3, q_1, q_2)}. \tag{3.88}$$

Here and further in this section, the notation $\varepsilon_{ik}$ is applied to denote only the Hermitian part of the corresponding components of the plasma dielectric permittivity tensor, $\Delta(k_3, q_1, q_2) = k_1^2[\varepsilon_{11}(q_1, q_2) + 2B(q_1, q_2)] + k_3^2\varepsilon_{33}(q_1, q_2)$. Their anti-Hermitian parts are omitted here, because they only describe kinetic damping of these TM-modes, which is out of scope of the investigations carried out in this section.

In the approximation mentioned above, specific expressions for the components of the dielectric permittivity tensor of the considered gyrotropic plasma $\varepsilon_{ik}(q_1, q_2) = \delta_{ik} + 4\pi i\sigma_{ik}(q_1, q_2)/\omega(q_1 + q_2)$, which are included in $\Delta(k_3, q_1, q_2)$, can be obtained using the expressions of the components of the electric current density $j_i(k_3, q_1, q_2) = \sigma_{ik}(q_1, q_2)E_k(k_3, q_1, q_2)$, which are given in Formulas (3.86) and (3.87).

One has to take into account that just the magnitude of the magnetic field of the SCTM-mode at the plasma interface but not its Fourier coefficient is written in the numerator on the right-hand side of expression (3.88). Thus, while carrying out the inverse Fourier transform, this numerator can be taken in front of the integral. Doing that in the approximate expressions for the tensors $\sigma_{ik}$ and $\varepsilon_{ik}$, one can neglect the ion contributions, because their values are at least $m_e/m_i$ times smaller than the corresponding electron contributions.

To derive an infinite set of equations for determination of the harmonics of the tangential component of the electric field of SCTM-modes on the plasma interface one should apply the following boundary conditions. First, the tangential component of the electric field of this TM-mode at the metal-dielectric interface is equal to zero. Second, the tangential component of the electric field of the SCTM-mode is continuous at the boundary of the plasma-dielectric interface. Third, there is the discontinuity of the tangential component of the magnetic field of SCTM-modes on the plasma-dielectric interface. This discontinuity exists due to the nonlinear electric current flowing along the plasma surface. This should be presented in explicit form since it differs from the corresponding expression obtained in the previous section in the case of a monochromatic RF field:

$$H_y(n_1, n_2)|_{z=+0}^{z=-0} = \sum_{\alpha} \frac{\Omega_\alpha^2}{ic|k_1|} \sum_{\substack{s,m1,m2,p1,p2;\\ p1\neq 0,\, p2\neq 0}} \frac{I_s(y_\alpha)\exp(i\beta_1 p_1)s\omega_\alpha G(m_1, m_2, p_1, p_2)}{\exp(y_\alpha)[\omega(n_1+m_1, n_2+m_2) - s\omega_\alpha]^2}$$
$$\times \sqrt{(\varepsilon_{33} + 2B)\varepsilon_{11}}\, E_x(n_1+m_1, n_2+m_2)|_{z=+0}, \tag{3.89}$$

where $\varepsilon_{ik}$ are only the Hermitian parts of the dielectric permittivity tensor of the studied gyrotropic plasma, as it was already mentioned above. The sums, which are written on the right-hand side of Formula (3.89), can be calculated independently over all the indices.

After application of the inverse Fourier transform to the right-hand side of expression (3.88), one can replace the tangential magnetic field $H_y(n_1, n_2)|_{z=-0}$ of this

TM-mode by the tangential component of its electric field calculated at the interface dielectric-plasma and the expression of the impedance of the dielectric layer using the nonlinear boundary conditions (3.89). All these operations allow one to derive an infinite set of equations for the Fourier harmonics of the tangential electric field on the plasma surface in the following form:

$$D(n_1, n_2)E_x(n_1, n_2)|_{z=0} - \sum_{\substack{p1,p2=-\infty \\ p1,p2\neq 0}}^{+\infty} F(n_1 + p_1, n_2 + p_2)E_x(n_1 + p_1, n_2 + p_2)|_{z=0} = 0, \qquad (3.90)$$

where

$$D(n_1, n_2) = 1 - \frac{\varepsilon_{\mathrm{d}} \coth(|k_1 a_{\mathrm{d}}|)}{\sqrt{(\varepsilon_{33} + 2B)\varepsilon_{11}}}, \qquad (3.91)$$

$$F(n_1 + p_1, n_2 + p_2) = \sum_{s,m1,m2=-\infty}^{+\infty} \frac{s\omega_e \Omega_e^2 \exp(in_1\beta_1) I_s(y_e) G(m_1, m_2, p_1, p_2)}{\exp(y_e)\omega(n_1, n_2)[\omega(n_1 + m_1, n_2 + m_2) - s\omega_e]^2}. \qquad (3.92)$$

Searching for solutions of the set of equations (3.90) becomes a complicated problem in comparison to the case that is studied in the previous Sect. 3.5. This is connected with the following circumstance. The features of the determinant factors in Formula (3.90) nearby the Fourier harmonics of the tangential electric field of this mode are very complicated. Namely, the numbers of both rows and columns of the determinant are defined just by pairs of numbers, for example, if for the rows there is $i = \{n_1, n_2\}$ then for the columns there is $j = \{p_1, p_2\}$.

The problem can be significantly simplified if the ratio of the operating frequencies of the pumping RF field is a ratio of natural numbers, i.e. if condition (3.80) is satisfied. Thus, the further considerations of this problem is carried out here under the condition of the following resonance:

$$\omega = \omega_{\mathrm{eig}} + \gamma, \qquad (3.93)$$

where $\omega_{\mathrm{eig}} = |s\omega_e| + \Delta_{Te} - N_0\omega_0$, $s$ is the number of the studied electron cyclotron harmonic and $N_0$ is a natural number whose value is determined by Formula (3.81).

The analysis of the resonance condition (3.93) confirms the convenience of choosing this new variable, namely, the transition to an effective value of the frequency of the external RF field. Then from (3.90) one can derive the approximate equation of the following type:

$$\gamma \frac{\mathrm{d}D(N_0)}{\mathrm{d}\omega_{\mathrm{eig}}} - F(N_0; +1)F(N_0 + 1; -1) - F(N_0; -1)F(N_0 - 1; +1) \approx 0, \qquad (3.94)$$

where

$$\frac{\mathrm{d}D(N_0)}{\mathrm{d}\omega_{\mathrm{eig}}} \approx -\sum_{s=1}^{+\infty} \frac{s\Omega_e^2 I_s^2(y_e)\exp(-2y_e)}{|\omega_e| y_e \Delta_{Te}^4 \varepsilon_{\mathrm{d}} \coth(|k_1 a_{\mathrm{d}}|)}. \tag{3.95}$$

In (3.95) $\Delta_{Te}$ is the shift of the electron SCTM-mode frequency from the cyclotron harmonic resonance of the studied TM-mode caused by thermal motion of the charged plasma particles: $\omega = s|\omega_e| + \Delta_{Te}$. The magnitude of this shift can be determined from the dispersion relations for the frequency of electron SCTM-modes, which are calculated in Sect. 3.1. Their expressions for the ranges of long and short wavelengths, respectively, are as follows:

$$\Delta_{Te} \approx -s_0|\omega_e|\frac{s_0^2 - 1}{2s_0!}(y_e/2)^{s_0-1}, \quad s_0 \geq 2, \quad y_e \ll 1; \tag{3.96}$$

$$\Delta_{Te} \approx s_0|\omega_e|\frac{\Omega_e^2 y_e^{-3/2}}{\sqrt{2\pi}\,\omega_e^2}, \quad y_e \gg 1. \tag{3.97}$$

Carrying out the numerical calculations of equations (3.90) one should take into account a sufficient number of terms over the indices $m_{1,2}$ and $p_{1,2}$ in the expressions for the $F(i, j)$ coefficients, respectively, to get the desired accuracy of the calculations. For analytical investigation, it will be sufficient to take into account only basic terms in the expressions for the $F(i, j)$, which have small resonant denominators.

Here the expressions of those coefficients $F(i, j)$ in Formula (3.94) are presented, which one can apply for analytical investigation of parametric instability of the electron SCTM-modes:

$$F(N_0; \pm 1) \approx \sum_{s=1}^{+\infty} \sum_{m1,m2=-\infty}^{+\infty} \frac{4s\Omega_e^2|\omega_e|[\omega(N_0) + m_1\omega_{01} + m_2\omega_{02}]I_s(y_e)}{[(\omega(N_0) + m_1\omega_{01} + m_2\omega_{02})^2 - s^2\omega_e^2]^2 \exp(y_e)} \times \left[J_{m1}^2(a_{E1})J_{m2}(a_{E2})J_{m2\mp1}(a_{E2}) - J_{m2}^2(a_{E2})J_{m1}(a_{E1})J_{m1\mp1}(a_{E1})\right] \tag{3.98}$$

$$F(N_0 \pm 1; \mp 1) \approx \sum_{s=1}^{+\infty} \sum_{m1,m2=-\infty}^{+\infty} \frac{4s^2\Omega_e^2\omega_e^2[\omega(N_0 \pm 1) + m_1\omega_{01} + m_2\omega_{02}]I_s(y_e)\exp(-y_e)}{[(\omega(N_0 \pm 1) + m_1\omega_{01} + m_2\omega_{02})^2 - s^2\omega_e^2]^2\omega(N_0 \pm 1)} \times \left[J_{m1}^2(a_{E1})J_{m2}(a_{E2})J_{m2\pm1}(a_{E2}) + J_{m2}^2(a_{E2})J_{m1}(a_{E1})J_{m1\pm1}(a_{E1})\right] \tag{3.99}$$

Even the first glance on the expressions (3.98) and (3.99) provides the following conclusion. Since the Bessel functions of the first kind [20] have an oscillating non-monotonic behavior depending on their argument and order, solutions of equation (3.94) have a very complicated dependence on the parameters of this RF field. That is why one can analytically examine the dependence of the instability growth rates on the external conditions only in some limiting cases.

The first of them is as follows. If any frequency of the components of this non-monochromatic RF field is not close to the electron cyclotron frequency, one can obtain an approximate cubic equation for the correction $\gamma$ to the eigenfrequency of the electron SCTM-modes. From (3.94), one can derive the following approximate expression for $\gamma$ under the condition of satisfying the resonance condition (3.93):

$$\begin{aligned}\gamma^3 &\approx 4s_0 y_e \left|\omega_e^3\right| \Delta_{Te}^4 \varepsilon_{\rm d} \coth(|k_1 a_{\rm d}|)\\ &\times \left[J_1(a_{E1})J_0(a_{E1})J_0^2(a_{E2}) + J_1(a_{E2})J_0(a_{E2})J_0^2(a_{E1})\right]\\ &\times \left\{ \sum_{m1,m2=-\infty}^{+\infty} \frac{\omega(N_0-1)+m_1\omega_{01}+m_2\omega_{02}}{[(\omega(N_0-1)+m_1\omega_{01}+m_2\omega_{02})^2 - s_0^2\omega_e^2]^2\omega(N_0-1)} \right.\\ &\times \left[J_{m1-1}(a_{E1})J_{m1}(a_{E1})J_{m2}^2(a_{E2}) + J_{m2-1}(a_{E2})J_{m2}(a_{E2})J_{m1}^2(a_{E1})\right]\\ &- \sum_{m1,m2=-\infty}^{+\infty} \frac{\omega(N_0+1)+m_1\omega_{01}+m_2\omega_{02}}{[(\omega(N_0-1)+m_1\omega_{01}+m_2\omega_{02})^2 - s_0^2\omega_e^2]^2\omega(N_0+1)}\\ &\times \left.\left[J_{m1}^2(a_{E1})J_{m2}(a_{E2})J_{m2+1}(a_{E2}) + J_{m2}^2(a_{E2})J_{m1}(a_{E1})J_{m1+1}(a_{E1})\right]\right\}.\end{aligned} \quad (3.100)$$

In the following the approximate expression (3.100) under the condition that the amplitudes of both external RF fields are sufficiently small, $a_{E1}, a_{E2} \ll 1$ is analyzed. This analysis shows that the growth rate of parametric instabilities of electron SCTM-modes is lower in this case in comparison with the case of a monochromatic RF pumping field, when the instability of these TM-modes is characterized by the following dependence of the growth rate on the pumping wave amplitude: $\mathrm{Im}\gamma \propto a_E^{2/5}$. The maximum magnitude of the growth rate in the present case can be determined approximately as follows:

$$\mathrm{Im}\gamma \approx \frac{\sqrt{3}}{2}\Delta_{Te}\left[\frac{y_e \Delta_{Te}\varepsilon_{\rm d}\coth(|k_1 a_{\rm d}|)}{s_0^3|\omega_e|}\right]^{1/3}(a_{E1}+a_{E2})^{2/3}. \quad (3.101)$$

However, if the amplitude of one of the pumping fields, for example, the first one is small, $a_{E1} \ll 1$, but the amplitude of the second field is not small, $a_{E2} \geq 1$, the expression (3.101) cannot be used. In this case, the value of the growth rate $\mathrm{Im}\gamma$ depends strongly on the value of the first factor in the square brackets of equation (3.100). The first term in these brackets is positive and it can be estimated as the product $a_{E1}J_0^2(a_{E2})/2$. At the same time, the second term can be either negative or positive. Estimation of its order of magnitude allows one to make the conclusion that the approximate value of this second term is equal to $J_1(a_{E2})J_0(a_{E2})$. Therefore, it can change both its absolute value and its sign. That is why the increase of the amplitude of the second RF field can lead either to a certain increase of the SCTM-mode growth rate as compared with the $\mathrm{Im}\gamma$ determined by (3.101) or to a reduction of its magnitude.

In fact, the most interesting feature in the study of this case is another one, namely, the change of $a_{E2}$ can lead to the regime of suppression of the electron SCTM-mode parametric instability. This can be realized if the approximate relation $a_{E1} J_0(a_{E2})/2 \approx -J_1(a_{E2})$ is true. This means that the parametric instability growth rate becomes so small that even a weak damping of these modes, determined for example by collisions between plasma particles, will be sufficient to suppress the growth of the instability.

The second limiting case is realized for another condition, namely, if the frequency of one of the components of the pumping RF field, for example, $\omega_{01}$, is close to the electron cyclotron frequency and the frequency of the second component is far from $|\omega_e|$. Then the analysis of equation (3.94), that describes the parametric instability of these TM-modes, is much more complicated compared to the previous case. In this case the Cardano method allows one to obtain the following expression for the correction $\gamma$ to the frequency of the studied SCTM-modes under the resonance condition (3.93):

$$\begin{aligned}\gamma^3 &\approx \tfrac{|\omega_e|}{s_0} y_e \Delta_{Te}^4 \varepsilon_{\mathrm{d}} \coth(|k_1 a_{\mathrm{d}}|) \tfrac{I_{s_0+1}(y_e)}{I_{s_0}(y_e)} \\ &\times\left[J_1(a_{E1}) J_0(a_{E1}) J_0^2(a_{E2}) + J_1(a_{E2}) J_0(a_{E2}) J_0^2(a_{E1})\right] \\ &\times\left\{\tfrac{[|\omega_e|(s_0+1)-\omega_0]}{\omega(N_0-1)(\Delta_{Te}+\gamma)^2}[J_1(a_{E1}) J_0(a_{E1}) J_0^2(a_{E2}) - J_1(a_{E2}) J_0(a_{E2}) J_1^2(a_{E1})]\right. \\ &\left. -[J_1(a_{E1}) J_0(a_{E1}) J_0^2(a_{E2}) + J_1(a_{E2}) J_0(a_{E2}) J_1^2(a_{E1})] \tfrac{[|\omega_e|(s_0+1)+\omega_0]}{\omega(N_0+1)(\Delta_{Te}+\gamma)^2}\right\}.\end{aligned} \quad (3.102)$$

Analyzing expression (3.102) one can see that if, for example, the amplitude of the second component of the pumping RF field is equal to zero, $a_{E2} = 0$, and the amplitude of the first component is small, $a_{E1} \ll 1$, one can obtain an approximate expression of the electron SCTM-mode growth rate in the case of a monochromatic RF pumping field: $\mathrm{Im}\gamma \propto a_E^{2/5}$. If the amplitudes of both components of the RF field are small, one can derive from expression (3.102) that $\mathrm{Im}\gamma \propto (a_{EI}+a_{E2})^{2/5}$. But in this case, the presence of the second component of the RF field introduces the influence of the plasma spatial dispersion on the parametric instability of the electron SCTM-modes. Namely, the factor $I_{n+1}(y_e)/I_n(y_e)$ appears in the expression for the growth rate as compared with expression (3.63), that was derived in the case of a monochromatic RF field. Therefore, the approximate expression for the growth rate has the following form:

$$\mathrm{Im}\gamma \approx |\Delta_{Te}| \frac{I_{s+1}(y_e)}{I_s(y_e)} \left\{ \frac{y_e^2 \varepsilon_{\mathrm{d}}^{-1}}{I_s(y_e) 2(s^2-1)^2} \right\}^{1/5} (a_{E1} + a_{E2})^{2/5}. \quad (3.103)$$

If the amplitude of the first component of the pumping RF field, whose operating frequency is close to the electron cyclotron resonance frequency, is small, increasing amplitude $a_{E2}$ of the second frequency component, which is operating far from the electron cyclotron resonance, slightly changes the value of $F(N_0 \pm 1; \pm 1)$, but strongly affects the value of the coefficients of the central row of the determinant that is created by the multipliers $F(N_0; \pm 1) \propto a_{E1} J_0^2(a_{E2})/2 + J_1(a_{E2}) J_0(a_{E2})$ located nearby the Fourier harmonics of the tangential electric field of electron SCTM-modes [see (3.90)]. Therefore, the value of Im$\gamma$ can either increase or decrease as compared to the case of monochromatic pumping RF field.

The case of arbitrary values of the amplitudes of plasma waves in an external RF field should be analyzed by numerical methods for specific values of the amplitudes $a_{E1}$ and $a_{E2}$ of the components of this field as well as of the ratio $\omega_{01}/\omega_{02}$ of their operating frequencies. One can expect without detailed investigation that:

- first, the value of the growth rate of the electron SCTM-mode parametric instability is a non-monotonous function of the parameters of the applied RF field;
- second, it is possible to suppress the parametric instability of these TM-modes by changing the parameters of the second component of the applied RF field, whose operating frequency is far from the electron cyclotron resonance.

## 3.7 Gas Discharges Sustained by Surface Electron Cyclotron TM-Modes

This section presents the results of developing the theoretical electrodynamic model of a gas discharge whose operation principle is based on the propagation of surface electron cyclotron TM-modes in planar metal waveguides with dielectric coating of their inner interfaces. It is shown that SCTM-modes can sustain the gas discharge in such a waveguide structure that is immersed into an external static magnetic field directed perpendicularly to the plasma interfaces. These waves are eigenmodes of such waveguide structures.

It is assumed that the gas discharge is sustained by volumetric ionization processes which happen due to collisions between electrons and gas particles filling the waveguide. Applying the analytical expressions obtained in Sects. 3.1 and 3.2 on the spatial distribution of the electron SCTM-modes it is possible to calculate the value of their energy flux and amount of their power that can be absorbed by the plasma during their propagation. This allows one to construct the power balance equation. One can further use this equation to investigate such discharge features as spatial distribution of plasma density that can be maintained in such a discharge, discharge length (this parameter allows one to estimate the length of the uniform plasma region that can be produced in this discharge), dependence of the parameters of the gas discharge on the strength of the applied constant external magnetic field and wavelength of the utilized electron SCTM-modes [48–51].

The geometry of the planar waveguide structure that can be employed for confining the gas discharges which are sustained by propagation of electron SCTM-modes is as follows. The region $z \leq -a_{\rm d}$ is occupied by the plane metal wall (see Fig. 3.22). It is assumed that this wall has a perfect electrical conductivity and is coated with a protective dielectric layer with dielectric constant $\varepsilon_{\rm d}$, whose thickness is $a_{\rm d}$. In the half space $z > 0$, the discharge chamber is filled with plasma. The applied external

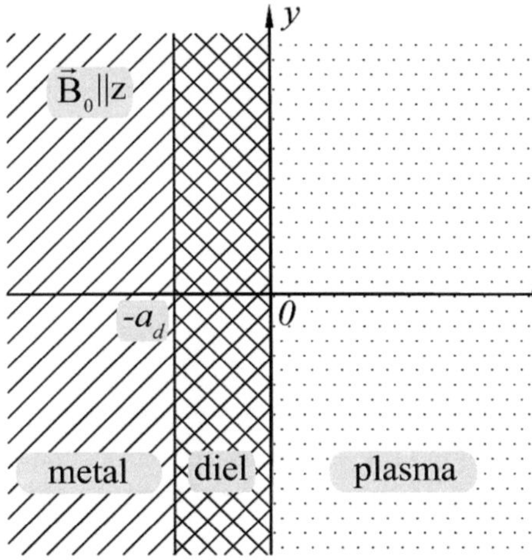


**Fig. 3.22** Schematic of the problem "gas discharges sustained by SCTM-modes"

static magnetic field $\vec{B}_0$ is directed perpendicularly to the metal wall, so that the direction of $\vec{B}_0$ coincides with that of the $\vec{z}$ axis. Electron SCTM-modes propagate along the plasma-dielectric interface (along the direction of the $\vec{x}$ axis) across the external magnetic field $\vec{B}_0$. The considered discharge structure is assumed to be homogeneous along the $\vec{y}$ axis. The motion of the plasma particles is described by the kinetic Vlasov-Boltzmann equation because, as it was shown in Chap. 2, the hydrodynamic theory does not allow one to describe correctly electromagnetic waves in the range of the harmonics of both the ion and electron cyclotron frequencies.

Since electron SCTM-modes are eigenwaves of the considered waveguide structure, they can be efficiently excited by external sources of electromagnetic energy. In addition to this fact one has to keep in mind that SCTM-modes are slow waves, that is why they can interact with the plasma particles more efficiently compared with bulk waves, which as a rule are fast waves. It is assumed that the SCTM-mode fields vary weakly along the direction perpendicular to the plasma-dielectric interface. This assumption means that the spatial dispersion of the plasma maintained in the discharge is weak along the direction of the $\vec{z}$ coordinate axis. This is consistent with the results of experimental measurements, which are presented in [52–55].

Here one considers the case when the applied external static magnetic field is not very strong, so that the following inequality is valid for the squared values of electron cyclotron and electron plasma frequencies: $\omega_e^2 < \Omega_e^2$. From one point of view, this condition is necessary for the existence of electron SCTM-modes in such waveguide structures. From the other point of view, in modern plasma technology devices, one should try to utilize a weak external static magnetic field in order to reduce the cost of the discharge plasma. This condition is typical for today's experiments, when the plasma obtained in a gas discharge is characterized by high density, and at the same time, the applied magnetic field is rather small. In addition, at present, one of the most important problems for practical utilization of any gas discharge is the choice of the type of the discharge, which allows one to create a plasma volume with a high density of charged particles [55].

The dependence of all the variables of the problem on the coordinates and time is assumed to have the following form:

$$A(x, z, t) = A_0(x, z)\exp(ik_1x - i\omega t). \tag{3.104}$$

Here the factor $A_0$ weakly depends on the x-coordinate, so that the inequality $A_0^{-1}\partial A_0/\partial x \ll k_1(x)$ is true. This approximation means that the values of the gas discharge parameters change slowly along the axis of the discharge chamber at distances which are of the order of the wavelength. One also assumes that the penetration depth of the electron SCTM-modes into the plasma is much larger than their wavelength (see Sect. 3.1) and that the ionization processes in the gas discharge occur due to collisions between electrons and neutral gas atoms that fill the discharge chamber. These collisions are described in terms of the effective frequency $\nu$ of momentum transfer. The value of this collision frequency is assumed to be much lower than the eigenfrequency of the electron SCTM-modes ($\nu \ll \omega$).

The spatial structure of the gas discharge is described by the local dispersion relation for electron SCTM-modes, the energy balance equation, and the ratio between the energy absorbed per unit length of the discharge and the local value of the plasma density [51]. The latter condition describes the selected type of gas discharge and primarily is determined by the operating regime of the conducted experiment. To determine the density of the produced plasma with the aid of the energy balance equation, one has to find the analytical expression for the electromagnetic field of these modes. Applying the Fourier analysis one can derive the analytical expressions for the components of the electromagnetic field of the electron SCTM-modes in the plasma and dielectric regions. In the coordinate space that is occupied by the dielectric layer $-a_d < z < 0$, the components of the electromagnetic field of SCTM-modes have the following form:

$$E_x^d(x, z, t) = E_x(0)\frac{\sinh[|k_1|(z + a_d)]}{\sinh(|k_1|a_d)}\exp(ik_1x - i\omega t),$$
$$H_y^d(x, z, t) = E_x(0)\frac{ik\varepsilon_d}{|k_1|}\frac{\cosh[|k_1|(z + a_d)]}{\sinh(|k_1|a_d)}\exp(ik_1x - i\omega t),$$
$$E_z^d(x, z, t) = -i\eta E_x(0)\frac{\cosh[|k_1|(z + a_d)]}{\sinh(|k_1|a_d)}\exp(ik_1x - i\omega t). \tag{3.105}$$

In the plasma region, $z > 0$, the components of the electron SCTM-mode fields are described by the following formulas:

$$E_x^{pl}(x, z, t) = F_0\exp(-k_\perp z)\exp(ik_1x - i\omega t),$$
$$H_y^{pl}(x, z, t) = F_0\big(ik/|k_1|\big)\sqrt{\varepsilon_{11}(\varepsilon_{33} + F_1)}\exp(-k_\perp z)\exp(ik_1x - i\omega t),$$
$$E_z^{pl}(x, z, t) = -ik_\perp F_0\exp(-k_\perp z)\exp(ik_1x - i\omega t), \tag{3.106}$$

where $F_0 = E_0[1 + \frac{i\,|k_1|\nu\omega\, h_e^4}{4\sqrt{\pi}\Omega_e^2 I_s^2(y_e)}\sqrt{(\varepsilon_{33}+F_1)\varepsilon_{11}}]$, $F_1 = 2k_1\varepsilon_{13}/|k_3|$, $\eta = k_1/|k_1|$, $k_\perp = \sqrt{\varepsilon_{11}k_{11}^2(\varepsilon_{33}+F_1)^{-1}}$, $\varepsilon_{ij}$ is the Hermitian part of the dielectric permittivity tensor of the plasma that is produced in the discharge, $E_0 = E_x(z=0)$ is the amplitude of the tangential component of the SCTM-modes calculated at the plasma interface, the penetration depth of these modes is equal to $\lambda_\perp = k_\perp^{-1}$, and $h_e = 1 - s|\omega_e|/\omega$.

The energy balance equation for the stationary regime of the gas discharge sustained by propagation of electron SCTM-modes can be written in the following form according to the conclusions of reference [51]:

$$\mathrm{d}S_x/\mathrm{d}x = -Q, \tag{3.107}$$

where $S_x$ is the average value of the wave energy flux density in the direction of the discharge chamber axis, and $Q$ is the average wave energy absorbed per unit length of the discharge plasma. The relationship between the normalized density of the produced plasma and the wave power absorbed per unit length of the discharge is specific for different types of discharges [53–56].

The average value of this wave energy flux density $S_x$ and the wave power $Q$ absorbed per unit length are presented in the form of the following integral expressions:

$$S_x = -\frac{c}{8\pi}\int_{-a_\mathrm{d}}^{\infty} \mathrm{Re}\big(H_y(z)E_z(z)\big)\mathrm{d}z, \tag{3.108}$$

$$Q = \frac{1}{2}\int_0^{\infty} \vec{j}(z)\vec{E}^*(z)\mathrm{d}z = Q_\mathrm{col} + Q_\mathrm{kin}, \tag{3.109}$$

where $j_n = \sigma_{nm}E_m$ is the Fourier coefficient of the electric current density, and $\sigma_{nm}$ is the plasma electric conductivity tensor. Its components have been calculated and presented in Sect. 2.2.

Applying the expressions (3.105) and (3.106) for the wave fields and those for the components of the $\sigma_{nm}$ tensor one can calculate the integrals (3.108) and (3.109) for the SCTM-mode energy flux density and the wave power absorbed per unit length of the discharge under the regime of Ohmic dissipation, which is characterized by the condition $Q_\mathrm{col} > Q_\mathrm{kin}$. The results of integration can be written in the following form [51]:

$$S_x = S_x^\mathrm{pl} + S_x^d \approx \frac{\omega \cdot E_0^2}{8\pi \cdot k_1 k_\perp}|\varepsilon_{11}|, \tag{3.110}$$

$$Q_\mathrm{col} \approx \frac{\nu \cdot \Omega_e^2 I_s(y_e)E_0^2 s^2}{8\pi \cdot h_e \cdot y_e \exp(y_e)k_\perp}(1 + \frac{2y_e k_\perp^2}{s^2 k_1^2 h_e}). \tag{3.111}$$

To get the result (3.111) one has to take into account only those terms in the components of the plasma conductivity tensor $\sigma_{nm}(\vec{k})$, which contain the effective collision frequency $\nu$ between the plasma electrons and neutral particles. Anti-Hermitian parts of the $\sigma_{nm}(\vec{k})$ tensor are neglected.

The part $S_x^{\text{pl}}$ of the electron SCTM-mode energy, which is flowing in the plasma region, is much larger than that part $S_x^d$ which is flowing in the dielectric region. Besides that, only the value $S_x^{\text{pl}}$ of the TM-mode energy flux density determines the intensity of ionization of neutral gas in the discharge chamber. Therefore, the value of $S_x^d$ can be ignored in the energy balance equation (3.107).

From the energy balance equation, one can find the expression for the plasma density profile along the electron SCTM-mode propagation direction in linear approximation:

$$N \approx N_0(1 - x/L), \tag{3.112}$$

where $N_0 = N(x = 0)$ is the plasma particle density nearby the location where the electromagnetic energy enters into the plasma from the antenna utilized for excitation of the operating eigenmode. This approximate dependence (3.112) of the produced plasma density on the $x$ coordinate is applicable for the regime of Ohmic heating if the following inequality is valid: $\nu \ll \omega$. The effective length of the discharge $L_{\text{col}}$ under the regime of electron SCTM-mode energy transfer into the produced plasma through the Ohmic channel of dissipation is determined by the following formula:

$$L_{\text{col}} = |\omega\, h_e/(\nu\, k_1)|. \tag{3.113}$$

If the pressure of the working gas is low, the frequency of the plasma electron collisions with neutral particles is significantly reduced, and the main mechanism of energy transfer of the electron SCTM-modes into the produced plasma is kinetic damping caused by the interaction of plasma particles with its interface. The value of the SCTM-mode energy, which is absorbed by the produced plasma, averaged over the period of these modes is determined by the anti-Hermitian part of the plasma conductivity tensor in this case. Unlike the case of bulk waves, its value is not exponentially small [17, 18]. Under the considered conditions the value of $Q_{\text{kin}}$ is larger than the value of $Q_{\text{col}}$ in both ranges of wavelengths (i.e. when it is larger or smaller than the Larmor radius of the plasma electrons).

Thus, the following expression is derived for the electron cyclotron frequency harmonics with numbers $s \geq 2$ in the long-wavelength range ($k_1^2 \rho_e^2 \ll 1$):

$$Q_{\text{kin}} \approx \frac{\omega \cdot \Omega_e^2 E_0^2 (s^2 - 1)^2 (k_1^2 \rho_e^2)^{(2S-3)}}{2\pi \exp(-\varsigma) k_\perp \omega_e^2 16^S s!^2}, \tag{3.114}$$

where $\varsigma = (\omega h_e \lambda_\perp / v)^2 < 1$.

One can derive the following analytical expression for the value of the electron SCTM-mode energy that is absorbed through the kinetic mechanism, in the range of short wavelengths ($k_1^2\rho_e^2 \gg 1$):

$$Q_{\text{kin}} \approx \frac{2\omega \cdot E_0^2 \Omega_e^6 \exp(\varsigma)}{\pi^{5/2} k_\perp (k_1 \rho_e)^9}. \tag{3.115}$$

Analyzing both obtained expressions (3.114) and (3.115) one can conclude that the value of the absorbed energy $Q_{\text{kin}}$ is directly proportional to the utilized operating frequency and their penetration depth into the produced plasma. This energy also increases with increasing plasma density. In the range of long wavelengths its value increases with increasing mode wavelength, but in the range of short wavelengths it increases with shortening of the wavelength. Comparing the expressions (3.111), (3.114) and (3.115) one can see that the regime of Ohmic dissipation of the electron SCTM-mode energy can be realized in the range of intermediate values of their wavelengths ($k_1\rho_e \sim 1$), when the inequality $\varsigma < 1$ is satisfied.

The length $L_{\text{kin}}$ of the gas discharge sustained due to kinetic damping of electron SCTM-modes is a little bit longer or of the same order of the electron Larmor radius. The peculiarity of this discharge is that its length $L_{\text{kin}}$ is shorter than the length $L_{\text{col}}$ of the gas sustained by collisional damping of these TM-modes or it is of the same order. This peculiarity is explained by the strong attenuation of these modes due to resonant interaction of plasma electrons gyrating along the Larmor orbits in the planes, which are parallel to the produced plasma interface. Another reason is that in the range of long and short wavelengths the group velocity of electron SCTM-modes is small compared to the case of the modes, which belong to the intermediate range of wavelengths. Hence, the Ohmic damping of these modes is significant just in this case.

Another parameter that determines the plasma volume maintained in the gas discharge by SCTM-modes is their penetration depth $\lambda_\perp$ into the plasma. As it was shown in Sect. 3.1, its value is larger than the electron Larmor radius $\rho_e$ in the intermediate wavelength range ($k_1\rho_e \leq 1$) and significantly longer than $\rho_e$ in the other two wavelength ranges, namely: short ($k_1\rho_e \ll 1$) and long ($k_1\rho_e \gg 1$) wavelengths. In the long wavelength range the value of the parameter $\lambda_\perp$ approaches the value $k_1^{-1}$, gradually with increasing mode wavelength. In contrast to this, in the range of short wavelengths ($k_1\rho_e \ll 1$) its value stays to be much larger than $2\pi \cdot k_1^{-1}$. The value of the ratio $\lambda_\perp \rho_e^{-1}$ increases with the wavelength of the SCTM-modes in the range of long wavelength and consequently decreases with the wavelength of these TM-modes in the range of short wavelengths, as well as with increasing value of the external static magnetic field. Analysis of the expression for the energy flux density of these modes indicates that its value increases with the frequency, the wavelength and the penetration depth of the electron SCTM-modes into the plasma.

Thus, the results of the present theoretical studies prove that it is possible to maintain a gas discharge in a planar metal waveguide with dielectric coating of its inner surface via propagation of electron SCTM-modes. It is theoretically predicted that under the regime of low working gas pressure, TM-modes transfer their energy to the plasma due to the mechanism of kinetic damping. In this case the discharge length along the propagation direction of these modes is short (a little bit larger than $\rho_e$), and the transverse size of the discharge is sufficiently large, namely, about 100 $\rho_e$. In a high-pressure working gas regime with intermediate range of the SCTM-mode wavelengths ($k_1\rho_e$ ~1) one can realize the situation, when the main mechanism of energy transfer of these TM-modes into the sustained plasma volume is Ohmic dissipation. Change of the wave energy transfer channel leads to complete changing of the scale of this gas discharge. This means that the length of the discharge can be about a hundred wavelengths of these TM-modes and the transverse size of the discharge reduces almost to several $\rho_e$ .

Analyzing the expression (3.113) for the discharge length and the formulas describing the dispersion of these TM-modes (see Sect. 2.1), one can conclude that the effective length of the gas discharge sustained by electron SCTM-modes in the regime of Ohmic dissipation increases with increasing thickness of the dielectric coating of the metal waveguide. In the case of the low-pressure working gas regime of this discharge, the Ohmic dissipation becomes ineffective due to the reduced collision frequency between the plasma particles since the value of this frequency decreases proportionally to the working gas pressure. Then the kinetic damping of these modes becomes the main channel of energy transfer. Analyzing the expressions (3.114) and (3.115) describing the mean value of electron SCTM-mode power absorbed through the kinetic channel of energy dissipation and the expression (3.111) obtained for the collisional channel (Ohmic dissipation), one can see that the inequality $Q_{\rm kin} > Q_{\rm col}$ is satisfied for both the short ($k_1\rho_e << 1$) and long ($k_1\rho_e$ ~ 1) wavelength ranges. Therefore, the possible range, where the opposite inequality can be realized, is the region of intermediate wavelengths, where $k_1\rho_e$ ~ 1. This peculiarity can be explained by the fact that the maximum of the electron SCTM-mode amplitude is in the peripheral layer of the produced plasma, whose thickness is about the penetration depth $\lambda_\perp$ of these modes. As it was shown in Sect. 3.1, the value of $\lambda_\perp$ has its maximum just in the ranges of short and long wavelengths. Thus the kinetic interaction of these modes happens in the most effective way in these ranges of wavelengths.

One can estimate the sizes occupied by a relatively uniform plasma produced in the gas discharge sustained by electron SCTM-modes excited by a commercial magnetron generator with operating frequency 2.45 GHz. Assuming that the applied magnetic field is $B_0 \leq 500$ G and the gas temperature is $T_e \approx 1$ eV, one can find that the electron cyclotron frequency is $\omega_e \approx 8.8 \times 10^9$ s$^{-1}$ with an electron Larmor radius $\rho_e \approx 0.67 \times 10^{-4}$ m. Thus under the regime of low-pressure operating gas the region of the relatively uniform plasma sustained by these TM-modes would be as follows $L_{\rm kin} \approx 0.4$ mm (along the plasma interface, across an external static magnetic field) and $\lambda_\perp \approx 40$ mm (perpendicular to the plasma interface). Therefore, such discharges could be useful for precise processing of solids and improvement of magnetron-type gas discharges.

## 3.8 Conclusions

In this Chapter, the dispersion relation of surface electron cyclotron TM-modes in planar plasma waveguide structures is derived for the case when the external static magnetic field is oriented perpendicularly to the interface of the confined plasma. The plasma region is separated by a protective dielectric layer from a metal interface. It has been proved that reducing the thickness of the dielectric coating and increasing the value of its dielectric constant lead to a change of the dispersion of these waves in the range of long wavelengths as compared to the electron Larmor radius. Their dispersion curves converge to the minimum values of the corresponding electron cyclotron frequency harmonics.

Decreasing the difference between the frequencies of these TM-modes and the frequency of the corresponding electron cyclotron harmonics leads to strong damping of these modes. Reducing the thickness of the plasma layer and the plasma density also increases the values of the SCTM-mode damping rates. Modeling a plasma inhomogeneity by a set of thin layers of homogeneous plasmas allows one to make the conclusion that plasma inhomogeneity decreases the frequency of SCTM-modes as compared with the case of a homogeneous plasma. However, the influence of the inhomogeneity of the plasma density on the variation of the frequency of these TM-modes is weak, because their frequency is mainly determined by the magnitude of the applied external static magnetic field and by the ratio of the wavelength to the Larmor radius.

Results of theoretical investigation of the beam-plasma interaction, which leads to excitation of these TM-modes by a charged particle beam flowing over the plasma interface, are also presented in this chapter. Both models of semibounded plasma and a plasma layer with a finite thickness are considered. Since the particle beam density is supposed to be small compared to the plasma density, its influence is essential only for the cases of the Cherenkov resonance condition or the condition of the anomalous Doppler effect. It is proved that the growth rates of the electron SCTM-mode instability under the condition of the Cherenkov resonance are larger than under the condition of the anomalous Doppler effect. Decrease of the plasma layer thickness leads to increasing instability growth rates of these TM-modes.

It is shown that the electron SCTM-mode instability caused by beam-plasma interaction can be realized for two different regimes, namely, for a resonant beam-plasma instability and for a dissipative beam instability. The values of the growth rates of the SCTM-mode instability in the regime of resonant beam-plasma interaction are larger than the values of corresponding growth rates for the dissipative beam instability. In the case of bounded plasma systems, where energy dissipation happens due to different processes, the effective collision frequency, which models all possible dissipative processes, can be large. That is why the beam-plasma interaction, which is realized under the condition of sufficiently strong dissipation of the plasma energy, can lead to instability of the electron SCTM-mode propagation in such plasmas.

In this chapter, the theory of the initial stage of development of parametric instabilities of SCTM-modes at harmonics of the electron cyclotron frequency is developed in the approximation of weak dispersion along the normal to the plasma and small amplitude of the external pumping RF field. Calculations are performed for a model of planar plasma-dielectric-metal waveguide structure with finite transverse size, which is influenced by an external static magnetic field oriented perpendicularly to the plasma interface.

An infinite set of equations for the tangential components of the electric field of electron SCTM-modes excited in this waveguide structure has been derived. Its numerical analysis allows one to find values for the growth rates of parametric instabilities of electron SCTM-modes in the range of long, intermediate and short wavelengths as compared to the electron Larmor radius. The most effective way to excite these modes is utilization of an external RF field with a frequency close to the eigenfrequencies of the electron SCTM-modes. These parametric instability growth rates are larger for the short-wavelength than for the long-wavelength range. The magnitudes of the growth rates of electron SCTM-modes are proportional to the normalized amplitude of the pumping RF field. This fact distinguishes parametric instabilities of these surface modes from those of bulk electron cyclotron waves. It should be mentioned as well that the growth rates of electron SCTM-modes have a significant dependence on the thickness of the plasma. The growth rates of long-wavelength parametric instabilities are maximal when the thickness of the plasma layer is of the same order as the penetration depth of these mode fields into the plasma.

It has been shown theoretically that a non-monochromatic pumping RF field significantly affects the development of parametric instabilities of electron SCTM-modes. An RF field which is a superposition of two frequency components is studied in detail. If these frequencies are both far from the cyclotron frequency, an external RF field with small amplitudes influences excitation of electron SCTM-modes less as compared to the case of a monochromatic pumping electric field. While increasing their amplitudes one can even reach the saturation regime of this instability. However, if the frequency of one of the components of the pumping RF field is close to the cyclotron frequency, then, by changing the amplitude of the second frequency component, one can either enhance or stabilize the parametric instability of the SCTM-mode.

The possibility to sustain a gas discharge utilizing SCTM-modes at harmonics of the electron cyclotron frequency is proved theoretically. It has been found that in a regime of low-pressure working gas the power transfer of these TM-modes into the plasma is more effective due to the channel of kinetic damping. In this case the length of such a discharge (along the propagation direction of these modes) is of the order of some electron Larmor radii and its transverse dimension is several hundreds of the Larmor radius. In the regime of high working gas pressure for intermediate values of SCTM-mode wavelengths ($k_1 \rho_e \sim 1$) one can realize the situation where the main channel of power transfer of these modes into the plasma particles is Ohmic dissipation. Therefore, this type of discharge sustained by electron SCTM-modes can be used for precise surface treatment of solids.

## References


1. Akhiezer, A. I., Akhiezer, I. A., Polovin, R. V., Sitenko, A. G., & Stepanov, K. N. (1975). *Plasma electrodynamics*. Oxford: Pergamon Press.
2. Krall, N. A., & Trivelpiece, A. W. (1973). *Principles of plasma physics*. New-York: MacGraw-Hill.
3. Lominandze, D. G. (1981). *Cyclotron waves in plasma*. Oxford: Pergamon Press.
4. Gross, E. P. (1951). Plasma oscillations in a static magnetic field. *Physical Review, 82*(2), 232–242.
5. Sen, H. K. (1952). Solar "enhanced radiation" and plasma oscillations. *Physical Review, 88*(4), 816–822.
6. Sitenko, A. G., & Stepanov, K. N. (1957). On the oscillations of an electron plasma in a magnetic field. *Soviet Physics JETP, 4*(4), 512–520.
7. Bernstein, I. B. (1958). Waves in a plasma in a magnetic field. *Physical Review, 109*(1), 10–21.
8. Drummond, W. E., & Rosenbluth, M. N. (1962). Anomalous diffusion arising from micro instabilities in a plasma. *Physics of Fluids, 5*(12), 1507–1513.
9. Stix, T. H. (1990). Waves in plasmas: Highlight from the past and present. *Physics of Fluids B, 2*(8), 1729–1743.
10. Azbel', M. Ia., & Kaner, E. A. (1957). Theory of cyclotron resonance in metals. *Soviet Physics JETP, 5*(4), 730–744.
11. Kaner, E. A. (1958). Cyclotron resonance in plasma. *Soviet Physics JETP, 6*(2), 425–427.
12. Kaner, E. A. (1958). Theory of cyclotron resonance. *Soviet Physics JETP, 6*(6), 1135–1138.
13. Kaner, E. A., & Skobov, V. G. (1967). Electromagnetic waves in metals in a magnetic field. *Soviet Physics Uspekhi, 9*(4), 480–503.
14. Lifshits, I. M., Azbel, M. Ya., & Kaganov, M. I. (1971). *Electron theory of metals*. Moscow: Nauka. (in Russian).
15. Steele, M. C., & Vural, B. (1969). *Wave interactions in solid state plasma*. New-York: McGraw-Hill.
16. Vladimirov, V. V., Volkov, A. F., & Mejlikhov, E. Z. (1979). *Plasma of semiconductors*. Moscow: Atomizdat. (in Russian).
17. Kondratenko, A. N. (1979). *Penetration of a field into plasma*. Moscow: Atomizdat. (in Russian).
18. Kondratenko, A. N. (1985). *Surface and bulk waves in bounded plasma*. Moscow: Energoatomizdat. (in Russian).
19. Girka, V. O., Girka, I. O., Kondratenko, A. M., & Pavlenko, I. V. (1996). Surface electron cyclotron waves in the metallic waveguide structures with two component filling. *Contributions to Plasma Physics, 36*(6), 679–686.
20. Abramowitz, M., & Stegun, I. A. (1964). *Handbook of mathematical functions*. Series 55. New-York: National Bureau of Standards, Applied Mathematics.
21. Kondratenko, A. N. (1976). *Plasma waveguides*. Moscow: Atomizdat. (in Russian).
22. Girka, V. O., Girka, I. O., & Pavlenko, I. V. (1997). HF surface cyclotron waves in planar waveguide structures with non-uniform plasma filling. *Journal of Plasma Physics, 58*(part1), 31–39.
23. Beidler, C. D., Grieger, G., Harmeyer, E., et al. (1995). *Helias reactor studies*. Preprint of Max-Planck-Institut für Plasmaphysik, 85748 Garching bei München, FRG, *IPP 2/330*.
24. Akhiezer, A. I., & Fainberg, Y. B. (1949). On the interaction of a charged particle beam with electron plasma. *Doklady AN SSSR, 69*, 555–556 (in Russian). (Akhiezer, A. I., & Fainberg, Ya. B. (2008). On the interaction of a charged particle beam with electron plasma. *Ukrainian Journal of Physics, 53*, 87–88 (in English, Translated and reprinted from *Doklady Akademii Nauk SSSR, 69*, 555–556 (1949)).
25. Bohm, D., & Gross, E. P. (1949). Theory of plasma oscillation B. Excitation and damping of oscillations. *Physical Review, 75*(12), 1864–1876.
26. Lawson, J. D. (1977). *The physics of charged-particle beams*. Oxford: Pergamon Press.

27. Miller, R. B. (1982). *An introduction to the physics of intense charged particle beams*. New York: Plenum Press.
28. Beletsky, N. N., Bulgakov, A. A., Khankina, S. I., & Yakovenko, V. M. (1984). *Plasma instabilities and nonlinear phenomena in semiconductors*. Kyiv: Naukova Dumka. (in Russian).
29. Kondratenko, A. N., & Kuklin, V. M. (1988). *Basics of plasma electronics*. Moscow: Energoatomizdat. (in Russian).
30. Vucovic, S. (1986). *Surface waves in plasma and solids*. Singapore: World Scientific.
31. Platzmann, P. M., & Wolf, P. A. (1973). *Waves and interactions in solid state plasmas*. New York: Academic Press.
32. Ostrikov, K., Neyts, E. C., & Meyyappan, M. (2013). Plasma nanoscience: from nano-solids in plasmas to nano-plasmas in solids. *Advances in Physics, 62*(2), 113–224.
33. Silin, V. P. (1973). *Parametric influence of high power radiation on plasma*. Moscow: Nauka. (in Russian).
34. Tzoar, N. (1969). Parametric excitation in plasma in a magnetic field. *Physical Review, 178*(1), 356–362.
35. Aliev, Yu M, & Zunder, D. (1970). Parametric excitation of upper and lower hybrid resonances. *Soviet Physics JETP, 30*(4), 718–720.
36. Aliev, Y. M., & Silin, V. P. (1973). Parametric effect on a plasma of high intensity irradiation near electron cyclotron frequencies. *Soviet Physics Technical Physics, 17*(1), 1767–1752.
37. Kitsenko, A. B., Lominadze, D. G., & Stepanov, K. N. (1974). The parametric excitation of the electron cyclotron oscillations of a plasma located in an alternating electric field. *Soviet Physics JETP, 39*(2), 294–298.
38. Girka, V., & Iarko, V. (2014). Parametric instability of surface electron cyclotron TM-modes. *Progress in Electromagnetics Research M (PIER M), 34,* 39–46.
39. Margot, J., & Moisan, M. (1992). Surface wave sustained plasmas in static magnetic fields for study of ECR discharge mechanisms. In M. Moisan, J. Pelletier (Eds.), *Microwave excited plasmas*. Amsterdam: Elsevier.
40. Dine, S., Booth, J.-P., Curley, G. A., et al. (2005). A novel technique for plasma density measurement using surface-wave transmission spectra. *Plasma Sources Science and Technology, 14*(4), 777–786.
41. Maier, S. A. (2007). *Plasmonics: Fundamentals and Applications*. New-York: Springer.
42. Wong, A. Y., Divergilio, W. F., & Iizuka, K. (1979). Double resonance excitation in the ionosphere. *Physics Letters, 75A*(1, 2):144–148.
43. Akiyama, H., Hirose, S., & Takeda, S. (1977). Experiment of double resonance parametric process in a plasma. *Journal of the Physical Society of Japan, 42*(5), 1715–1720.
44. Milovich, J. L., Fried, B. D., & Morales, G. J. (1984). Growth rates of parametric instabilities driven by two pumps. *Physics of Fluids, 27*(7), 1647–1658.
45. Besson, T., & Edwards, W. S. (1996). Two-frequency parametric excitation of surface waves. *Physical Review E, 54*(1), 507–513.
46. Schvets, O. M., Dikiy, I. A., Kalinichenko, S. S., et al. (1986). Absorption of Alfven waves and plasma production in the Omega and Uragan-3 toroidal devices. *Nuclear Fusion, 26*(1), 23–41.
47. Korzh, A. F., & Stepanov, K. N. (1988). Parametric instability of plasma immersed into the field of two pumping waves. *Fizika Plazmy, 14*(6), 698–705. (in Russian).
48. Girka, V. O., Azarenkov, M. O., & Pavlenko, I. V. (1997). Model of plasma producer on surface cyclotron waves in metal waveguide with dielectric sheath. In *Contributed Papers of the 10th Joint Workshop on Electron Cyclotron Emission and Electron Cyclotron Resonance Heating, Netherlands* (pp. 621–628).
49. Girka, V. O. (1999). Plasma sources of magnetron type based on surface cyclotron waves propagation. In *Contributed Papers of the 26th EPS Conference on Controlled Fusion and Plasma Physics, Netherlands* (pp. 901–904).
50. Azarenkov, N., Girka, V., & Pavlenko, I. (2000). Comparative analysis of a low pressure gas discharges sustained by cyclotron surface waves. In *Proceedings of the 27th EPS Conference on Controlled Fusion and Plasma Physics, Hungary* (pp. 1469–1472).

51. Girka, V. O. (1999). Electron heating during microwave gas discharge sustained by surface cyclotron waves. *Physica Scripta, 60,* 257–260.
52. Moisan, M., & Zakrzewski, Z. (1986). Plasmas sustained by surface waves at microwave and RF frequencies: experimental investigation and applications. In J. M. Proud & L. H. Luessen (Eds.), *Radiative processes in discharge plasmas. NATO ASI, Series B* (Vol. 149, pp. 381–430).
53. Margot, J., & Moisan, M. (1991). Electromagnetic surface waves for a new approach to the investigation of plasmas produced at electron cyclotron resonance. *Journal of Physics D: Applied Physics, 24*(9), 1765–1788.
54. Wilhelm, R. (1993). ECR plasma sources. *NATO ASI, Series B: Physics, 302,* 161–180.
55. Aliev, Yu M, Schlüter, H., & Shivarova, A. (2000). *Guided-wave produced plasmas*. New York: Springer.
56. Aliev, Yu M, Boev, A. G., & Shivarova, A. P. (1982). On the non–linear theory of a long gas discharge produces by an ionizing slow electromagnetic wave. *Physics Letters, 92A,* 235–242.

# Chapter 4
# Surface Electron Cyclotron X-Mode Waves

Propagation of surface electron cyclotron waves was studied experimentally for the first time under the conditions of normal orientation of the external stationary magnetic field in respect to a semiconductor plasma interface [1–3]. The cyclotron resonance was observed therein with the help of magneto-optic methods of detecting the absorption of far-infrared laser radiation by a hetero-structure InAs–GaSb semiconductor. Very strong magnetic fields, $B_0 \geq 10$ T, and very low temperatures (of the order of liquid helium temperature) were applied in those experiments, which are analyzed in the monograph [4].

First results of studying surface electron cyclotron waves for the case when the external stationary magnetic field was oriented parallel to the plasma interface were discussed in the review [5]. When one studies magneto-plasma waves in semiconductors, this orientation of $\vec{B}_0$ is called Voigt geometry. The theory of electron cyclotron resonance in metals provided an efficient tool for experimentalists for studying Fermi surfaces in metals as well as different properties of the energy spectra of the conduction electrons (see [6–13] and references therein). After developing the theory of electron cyclotron resonance in metals, extensive theoretical investigation of surface electromagnetic waves had started both in the case of external stationary magnetic field $\vec{B}_0$ parallel to the interface of metals of different type [14–19], and in the case of arbitrary orientation of $\vec{B}_0$ to these interfaces [20]. Results of experimental studies of electron cyclotron waves in metal and semiconductor samples confirmed the general theoretical principles and clarified them in some details [21–23].

## 4.1 Surface Electron Cyclotron X-Mode Propagation Along Semibounded Plasmas–Dielectric Interface

Here the case of a plasma waveguide consisting of homogeneous plasma, which occupies the half-space $x > 0$, is considered, where the plane $x = 0$ is the boundary to a dielectric of infinite thickness and dielectric constant $\varepsilon_d$ (see Fig. 4.1). The



V. Girka et al., *Surface Electron Cyclotron Waves in Plasmas*,
Springer Series on Atomic, Optical, and Plasma Physics 107,
https://doi.org/10.1007/978-3-030-17115-5_4

plasma–dielectric boundary is assumed to be sufficiently sharp, so that the thickness of the transition layer is much less than the depth of surface wave penetration into the plasma. The external stationary magnetic field $\vec{B}_0$ is parallel to the plasma interface and oriented along the $\vec{z}$ coordinate axis. The electron surface cyclotron X-mode (SCX-mode) propagates exactly perpendicular to the magnetic field $\vec{B}_0$, which means along the plasma interface (along the $\vec{y}$ axis), and its amplitude decreases with going into the plasma core (along the $\vec{x}$ axis). Such mutual orientation of the plasma interface and external stationary magnetic field is typical for plasma magnetic confinement devices as tokamaks and stellarators.

The motion of plasma particles is described by the kinetic Vlasov–Boltzmann equation, where the unperturbed distribution function of the plasma particles is a Maxwellian. The diffusive model of particle reflection from the plasma interface is applied. In this case, the kinetic equation can be solved easier (see Chap. 2) as compared to the model of mirror reflection. It should be noted, that the simplest link between the Fourier images of electric current density and electric field can be obtained under the application of the model of backward reflection. However, its utilization needs an additional physical justification. Therefore, here the model of diffusive reflection is applied, which is often utilized in solving problems of plasma particle interaction in solids [24]. The space of the waveguide structure is assumed to be uniform along the $\vec{z}$ axis. Under these conditions, one can use the expressions for the components of the plasma dielectric permittivity tensor $\varepsilon_{ik}$, which are derived in Sect. 2.5. If one omits their anti-Hermitian parts, they are identical to the expressions for the tensor $\varepsilon_{ik}$ in the case of an infinite magnetoactive plasma [25–30].

The Maxwell equations are applied to describe the electromagnetic fields of SCX-modes. The dependence of SCX-mode fields on the longitudinal coordinate $y$ and time $t$ is chosen in the following form: $E, H \propto \exp(ik_2 y - i\omega t)$, where $k_2$ and $\omega$ are the component of the wave vector along the $\vec{y}$ axis and the angular electron SCX-mode frequency, respectively. The case of slow waves is considered, for

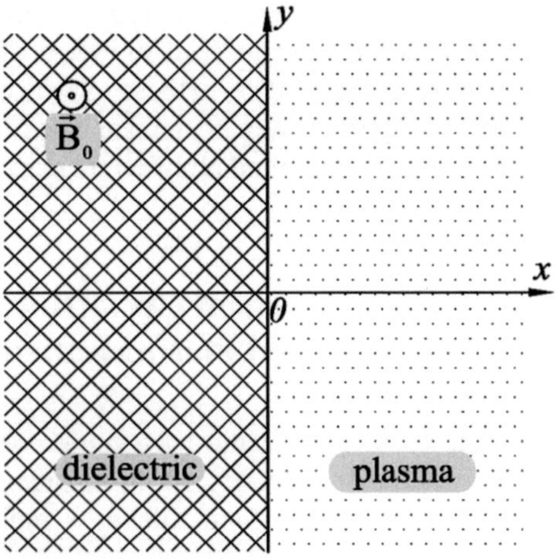


**Fig. 4.1** Schematic of the problem "semi-bounded plasmas–dielectric"

which $\upsilon_{\rm ph}^2|\varepsilon_{ik}| \ll c^2$, where $c$ and $\upsilon_{\rm ph} = \omega/k_2$ are the speed of light in vacuum and the SCX-mode phase velocity, respectively. The spatial dispersion of the plasma is assumed to be weak: $\left|\vec{k}\right|\upsilon_{T\alpha} \ll \omega - s|\omega_\alpha|$, where $\vec{k} = \vec{k}_1 + \vec{k}_2$ is the SCX-mode wave vector, which means that the waves under study propagate precisely perpendicular to the external magnetic field, that is oriented along the $\vec{z}$ axis.

Under these conditions, the set of Maxwell equations can be separated into two independent subsystems. One of them describes the X-mode with the components $E_x$, $E_y$, $H_z$ [31].

Fourier transform in respect to time and coordinate $y$ makes it possible to derive the following set of differential equations for the components of SCX-modes:

$$\begin{cases} ikH_z = \partial E_y/\partial x - ik_2 E_x; \\ ik_2 H_z = 4\pi\, j_x/c - ikE_x; \\ ikE_y = \partial H_z/\partial x + 4\pi\, j_y/c, \end{cases} \tag{4.1}$$

where $k = \omega/c$ is the wave vector in vacuum. Now, one has to analyze whether the SCX-mode field components are even or odd functions of the coordinate $x$. Changing the sign of the coordinate $x$ in these components in the set of equations (4.1) does not allow to answer this question. In this respect, this mode distinguishes from the TM-mode studied in Chap. 3. Because of the absence of any symmetry in the fields of the X-mode in respect to changing the sign of the coordinate $x$ to $(-x)$, one has to continue the SCX-mode fields by zero in carrying out the Fourier transform in respect to this coordinate. For demonstration of this, one may consider as an example the first equation from the set (4.1):

$$ikH_3 = \int\limits_0^{+\infty} \frac{\exp(-iqx)}{2\pi}\frac{\partial E_y}{\partial x}{\rm d}x - ik_2 E_1, \tag{4.2}$$

where the figures in the subscripts of the field symbols mean that there is the Fourier image but not the function of the real coordinate $x$. Thus, the Fourier transform gives the following set of three equations which includes the values of the tangential fields at the plasma interface:

$$\begin{cases} H_3 = \big(qE_2 - k_2E_1 + iE_y(0)/(2\pi)\big)/k; \\ H_3 = -\big(k/k_2\big)(\varepsilon_{11}E_1 + \varepsilon_{12}E_2); \\ H_3 = (k(\varepsilon_{22}E_2 - \varepsilon_{12}E_1) + H_z(0)/(2\pi i))/q, \end{cases} \tag{4.3}$$

where the expressions of the components of dielectric permittivity tensor $\varepsilon_{ik}$ are calculated in Chap. 2. Solving the set of algebraic equations (4.3) for the Fourier components of the SCX-mode fields, one can derive the following equation for the Fourier image of the tangential electric field component:

$$E_2\left(k^2k_2\varepsilon_{12}^2 - q^2k_2\varepsilon_{11} - k_2^3\varepsilon_{22} + k^2k_2\varepsilon_{11}\varepsilon_{22}\right)$$
$$= \frac{E_y(0)}{2\pi}ik_2(q\varepsilon_{11} - k_2\varepsilon_{12}) - \frac{iH_z(0)}{2\pi}k_2\left(k_2^2 - k^2\varepsilon_{11}\right). \tag{4.4}$$

The following two assumptions make it possible to simplify the expression (4.4). First, the X-modes studied here are slow waves, which means that $\left|k_2^2;\, q^2\right| \gg k^2|\varepsilon_{ik}|$. Second, since the spatial dispersion of the present plasma is weak, the following condition can be applied: $(k_2^2+q^2)\rho_\alpha^2/2 \ll 1$. These assumptions make it possible to use the following relation between the components of this plasma dielectric permittivity tensor: $\varepsilon_{11} \approx \varepsilon_{22}$.

It is a pity that one cannot derive the dispersion relation of SCX-modes for arbitrary harmonics of the electron cyclotron frequency, in contrast to the SCTM-modes studied in Chap. 3. That is why here the investigations are restricted to just the case of the second electron cyclotron harmonic. This restriction cannot be considered as a strong one, since excitation of just the second harmonic of the electron cyclotron frequency is most often utilized in fusion plasma experiments. The expressions for the two components of the tensor $\varepsilon_{ik}$, which take part in the expression (4.4), have the following approximate form with account for the assumptions mentioned above:

$$\varepsilon_{11} \approx \varepsilon_{22} \approx 1 - \frac{\Omega_e^2}{\omega^2 - \omega_e^2} - \frac{\Omega_e^2 s^2 (y_e/2)^{s-1}}{\omega 2s!(\omega - s|\omega_e|)}; \tag{4.5}$$

$$\varepsilon_{12} = -\varepsilon_{21} \approx \frac{i|\omega_e|\Omega_e^2}{\left(\omega^2 - \omega_e^2\right)\omega} + \frac{\Omega_e^2 s^2 (y_e/2)^{s-1}}{\omega 2s!(\omega - s|\omega_e|)}, \tag{4.6}$$

where $y_e = \rho_e^2(k_2^2 + q^2)/2$, and $\rho_e$ is the averaged value of the Larmor radius of the plasma electrons.

Application of the assumptions mentioned above makes it possible to derive the following approximate expression for the Fourier image of the tangential electric field, making its inverse Fourier transform:

$$\int_{-\infty}^{+\infty} E_2 e^{iqx}\mathrm{d}q = \int_{-\infty}^{+\infty} \frac{iE_y(0)(k_2\varepsilon_{12} - q\varepsilon_{11})}{2\pi\,\Delta(q)} e^{iqx}\mathrm{d}q + \int_{-\infty}^{+\infty} \frac{iH_z(0)k_2^2}{2\pi\,\Delta(q)} e^{iqx}\mathrm{d}q, \tag{4.7}$$

where $\Delta(q) = (q^2 + k_2^2)\varepsilon_{11}(q)$. The improper integrals in (4.7) can be calculated by the theory of residues. To apply this theory these integrals should be transformed into contour integrals by adding an infinite half-circle in the upper complex half-plane of the parameter $q$. Curvilinear integrals along this half-circle are equal to zero according to the Jordan's lemma since their integrand functions are approaching zero due to the factor $\exp(iqx)$ in which the parameter $q \to +i\infty$. On one hand, adding such half-circles does not change the values of the integrals in (4.7), whereas on the other hand, improper integrals with real variables transform into contour integrals in the upper complex $q$ half-plane.

According to the theory of residues, these integrals are equal to $2i\pi \sum_{j=0}^{N} res\left[F(q_j)\right]$ in the simplest case in which the denominator of the integrand function $F(q)$ has simple zeros [here the $q_j$ are roots of the denominator of $F(q)$]. In the present case of electron SCX-modes at the second harmonic, the equation $\Delta(q_j) = 0$ has two simple roots in the upper complex half-plane. One of them is called the kinetic root, since it satisfies the condition $\varepsilon_{11}(q_0) = 0$ in which the diagonal component of the permittivity tensor is written in kinetic approach. Then it is reasonable to call the second root $q_1$ as hydrodynamic root, since it is given by $q_1 = i|k_2|$.

Then the following expression can be easily derived from (4.7) for the plasma impedance $Z_{\text{pl}}(0) = E_y(0)/H_z(0)$:

$$Z_{\text{pl}}(0) = -\sum_{j=0}^{s-1}\left[\frac{\mathrm{d}\Delta(q)}{\mathrm{d}q}\right]_{q_j}^{-1}\left[\left(k_2\varepsilon_{12} - q_j\varepsilon_{11}\right)Z_{\text{pl}}(0) + k_2^2/k\right]. \tag{4.8}$$

Here, like in (4.5) and (4.6), the number of the electron cyclotron harmonic should be $s = 2$, as suggested above. It is clear from expression (4.5) for $\varepsilon_{11}$, that when studying the waves at the third electron cyclotron harmonic, there are two kinetic roots in the upper complex half-plane. In the case of waves at the fourth electron cyclotron harmonic there are three roots, and so on. This makes the problem of finding the roots more complicated, and therefore it is impossible to derive a dispersion relation for the general case (for arbitrary value of the harmonic number $s$).

The dispersion relation can be derived for a definite harmonic, setting the plasma surface impedance (4.8) equal to that of the dielectric surface. In this way, the dispersion relation for the second electron cyclotron harmonic can be written in the following form:

$$i\frac{k_2 q_0}{q_0^2 + k_2^2} = f(s), \tag{4.9}$$

where $f(s) = s-1$ for the waves which propagate along the $\vec{y}$ axis and $f(s) = -s-1$ for those which propagate in the opposite direction. Solutions of (4.9) can be written in the following form:

$$k_2 = \frac{1}{\rho_e}\sqrt{\frac{h_e}{3}}\begin{cases}\sqrt{5}-1, & k_2 > 0,\\ -\dfrac{\sqrt{37}+1}{3}, & k_2 < 0,\end{cases} \tag{4.10}$$

where $h_e = 1 - 2|\omega_e|/\omega$ is the relative shift of the wave frequency from the resonance.

As it follows from expression (4.10), the dispersion of SCX-modes is normal and their frequency is a little bit higher than $2|\omega_e|$ $(0 < h_e \ll 1)$. The depth of wave penetration into the plasma is of the order of the wavelength and much larger than the electron Larmor radius. The absolute value of the wavenumber $k_2$ is smaller for wave propagation along the $\vec{y}$ axis than under its propagation in the opposite direction.

This is explained by the anisotropy of the medium. There is a specific direction at the plasma interface, which is associated with the particles' Larmor gyration in the external stationary magnetic field. For electrons, the direction of gyration is opposite to that of the $\vec{y}$ axis.

SCX-mode damping takes place due to electron collisions with ions and the plasma interface. The first phenomenon is easy to be taken into account since the collision frequency $\nu_e$ explicitly is in the expression for the relative shift of the wave frequency from the resonance. The damping rate associated with the collision frequency $\nu_e$ can be defined via extracting the imaginary part from the fraction $ik_2 q_0/\left(q_0^2 + k_2^2\right)$. Doing that, one has to consider complex values $k_2 \to k_2' = k_2 + i\delta_\nu$, $h_e \to h_e' = h_e + i\nu_e/\omega$ and take into account that $|\delta_\nu| \ll |k_2|$, and $|\nu_e/\omega| \ll h_e$. Then one can derive the damping rate for the second electron cyclotron harmonic:

$$\delta_\nu = \frac{k_2 \nu_e}{2\omega h_e}. \tag{4.11}$$

Collisions between the particles do not relate in their nature to the orientation of the external magnetic field with respect to the plasma interface. That is why the damping rate has the same expression under the conditions of weak spatial dispersion for both cases of $\vec{B}_0$ orientation, perpendicular and parallel to the plasma interface.

To calculate the damping rate associated with electron collisions with the plasma interface (it is an analogue to Landau damping), one can get use of the SCX-mode damping rate $\delta_q$ caused by the dispersion of the medium. It has a complicated dependence on the wavenumber and the number of cyclotron harmonic. The order of magnitude is given by:

$$\delta_q \propto k_2 (h_e)^{1/(2s-2)}, \tag{4.12}$$

where $s$ is the number of cyclotron harmonic.

The total damping rate is $\delta = \delta_q + \delta_\nu$. It depends on the direction of wave propagation: $\delta(k_2 < 0) \neq \delta(k_2 > 0)$, due to the anisotropy of the medium. The proportionality factor in (4.12) drastically depends on the chosen model of plasma particle collisions with the interface. It is smaller for the model of mirror reflection, and larger for diffusive reflection.

## 4.2 Surface Electron Cyclotron X-Mode Propagation Along Plasma–Metal Interface

In this Section, the dispersion properties and damping of electron SCX-modes which propagate along the plasma–metal interface are studied for the case of the second harmonic electron cyclotron frequency. The propagation direction of these waves is shown to coincide with the direction of electron gyration in the external static magnetic field. The damping rates of these waves are given in [32, 33].

The model of semibounded uniform, completely ionized plasmas, which consist of ions of one species and electrons is applied. The coordinate system is chosen in such a way, that the plasma occupies the half-space $x > 0$, and the metal is located in the half-space $x \leq 0$ (see Fig. 4.2). The external stationary magnetic field $\vec{B}_0$ is directed along the $\vec{z}$ axis. SCX-modes propagate along the $\vec{y}$ axis, their fields do not depend on the $z$ coordinate. The diffusive model of plasma particle reflection from the boundary of the two media is applied. This model describes in the most proper way the interaction of particles in the plasma at the boundaries to solids [33]. The plasma interface is assumed to be sharp so that the thickness of the transient layer is much smaller than the electron Larmor radius $\rho_e$. The original equations which describe the dynamics of plasma particle motion and electromagnetic fields of this SW are the kinetic Vlasov–Boltzmann equation and the set of Maxwell equations. Plasma particles are assumed to be described by a Maxwellian equilibrium distribution function. The spatial plasma dispersion is assumed to be weak so that the following inequality is valid: $\left(\vec{k}\right)^2 \rho_e^2/2 \ll 1$. The phase velocity of the waves is assumed to be much lower than the speed of light in vacuum (slow waves). The external magnetic field should be sufficiently weak so that the following relation is valid: $\Omega_e^2 \gg \omega_e^2$.

The dependence of the wave field amplitudes on coordinates and time is given in the following form:

$$f(\vec{r}, t) = f(x) \exp(ik_2 y - i\omega t). \tag{4.13}$$

The Fourier images of the components of the electric plasma current can be found as result of solving the kinetic equation:

$$j_i(q) = \sigma_{ik}(q) E_k(q) + \frac{c}{4\pi} I_i(q). \tag{4.14}$$

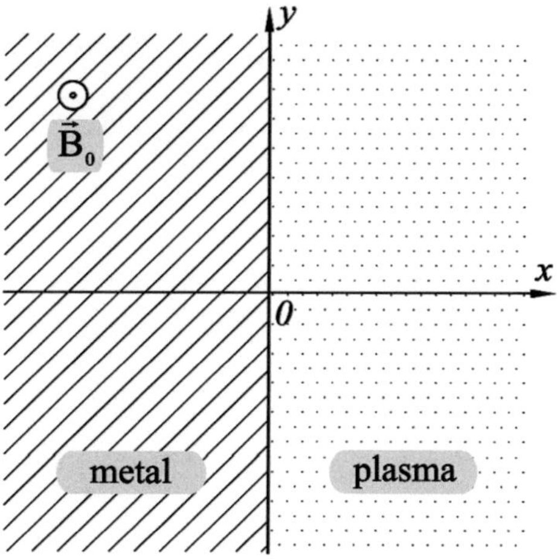


**Fig. 4.2** Schematic of the problem "plasma–metal"

In (4.14), $\sigma_{ik}$ is electric conductivity tensor of an infinite uniform plasma [25–27, 34, 35],

$$I_i(q) = \int_0^\infty \mathrm{d}q' Q_{ik}(q, q') E_k(q'), \tag{4.15}$$

where $Q_{ik}(q, q')$ is the integral part of the conductivity kernel caused by particle reflection from the plasma interface [24]. To calculate the integral (4.15) under the condition of weak plasma dispersion ($|q\rho_e| \ll 1$) it is sufficient just to consider the first nonzero term. For the diffusive model of particle reflection, one can then derive the following approximate expressions [24, 33]:

$$\begin{aligned}
Q_{11} &\approx \sum_\alpha L_\alpha\left(1 + 2\gamma_\alpha^2 - 9(1 - 6\gamma_\alpha^2)P_\alpha\right),\\
Q_{12} = -Q_{21} &\approx \sum_\alpha 3i\gamma_\alpha L_\alpha\left(1 + (7 + 8\gamma_\alpha^2)P_\alpha\right),\\
Q_{22} &\approx \sum_\alpha L_\alpha\left(2 + \gamma_\alpha^2 + 3(4\gamma_\alpha^2 + 5\gamma_\alpha^2 + 6)P_\alpha\right),
\end{aligned} \tag{4.16}$$

where $P_\alpha = \frac{1+\exp(2\pi i \gamma_\alpha)}{2(1-4\gamma_\alpha^2)(9-4\gamma_\alpha^2)}$, $L_\alpha = \frac{\upsilon_{T\alpha}\Omega_\alpha^2}{\pi^{3/2} c\omega_\alpha^2 (1-\gamma_\alpha^2)^2}$, $\gamma_\alpha = \frac{\omega + i\nu_\alpha}{\omega_\alpha}$, and $\upsilon_{T\alpha} = \sqrt{2T_\alpha / m_\alpha}$ is the thermal velocity of plasma particles of species $\alpha$.

Processing the Maxwell equations one can derive the set of equations which describe the extraordinary polarized $E$-wave with the field components $E_x$, $E_y$, $H_z$ or in other words, the X-mode:

$$\begin{cases}
iqH_3 = \dfrac{H_z(0)}{2\pi} + ikE_2 - \dfrac{4\pi}{c} j_2;\\
ikH_3 = -\dfrac{E_y(0)}{2\pi} + iqE_2 - ik_2E_1;\\
ik_2H_3 = -ikE_1 + \dfrac{4\pi}{c} j_1;
\end{cases} \tag{4.17}$$

where $E_y(0)$ and $H_z(0)$ are the fields of the studied mode at the plasma interface. As it follows from (4.17), the wave fields are not characterized by any symmetry. They are neither odd nor even functions of the coordinate $x$. That is why they are continued in the half-space $x < 0$ by zero. The set of Maxwell equations is solved by Fourier transformation. Then one can derive the following expression for the Fourier image of the tangential component of the X-mode electric field from the set (4.17):

$$\begin{aligned}
E_2 = \frac{1}{\Delta}\Bigg\{ & \frac{iE_y(0)}{2\pi}(k_2\varepsilon_{12} - q\varepsilon_{11}) - \frac{iH_z(0)}{2\pi k}(k^2\varepsilon_{11} - k_2^2)\\
& + iI_1\frac{k^2\varepsilon_{12} - k_2 q}{k} + iI_2\frac{k^2\varepsilon_{11} - k_2^2}{k}\Bigg\},
\end{aligned} \tag{4.18}$$

where $\Delta = (q^2 + k_2^2)\varepsilon_{11} - k^2(\varepsilon_{11}^2 + \varepsilon_{12}^2)$, $I_i(q) \approx Q_{ik}(q)E_k(0)$, and $\varepsilon_{ik} = 1 + 4\pi i\sigma_{ik}/\omega$ is the infinite plasma permittivity tensor [25–27, 34, 35]. Those terms on the right hand side of (4.18), which are proportional to $I_1$ and $I_2$, are caused by wave damping associated with particle reflection from the metal interface. That is why one can neglect them in studying the dispersion properties of SCX-modes.

To obtain the expression for the field $E_y(x)$ one has to integrate the (4.18) over $q$. Doing that one can neglect the terms proportional to the fraction $k^2k_2^{-2}$ in the integrand expressions since here the case of slow waves is under consideration. Then the following expression for the tangential component of the electric field of this mode can be derived:

$$E_y(x) = \sum_{n=0}^{s-1}\left(\frac{\partial\Delta}{\partial q_n}\right)^{-1}\left\{E_y(0)(q_n\varepsilon_{11} - k_2\varepsilon_{12}) - H_z(0)\frac{k_2^2}{k}\right\}\exp\{iq_nx\}, \quad (4.19)$$

where $q_n$ are solutions of the equation $\Delta(q) = \varepsilon_{11}(q)(k_2^2 + q^2) = 0$ in the upper complex half-plane.

The expressions for $\varepsilon_{ik}$ can be simplified under the condition of weak spatial dispersion, $\varepsilon_{11} \approx \varepsilon_{22}$. It is assumed that the SCX-mode propagates at the $s$-th harmonic of the electron cyclotron frequency such that $\omega = s|\omega_e|(1 - h_e)^{-1}$, where $h_e \ll 1$. Then one has to take into account the terms with $n = 0, \pm 1, s$ in the expressions for the components of $\varepsilon_{ik}$, and as the result they get the following form:

$$\varepsilon_{11} \approx \varepsilon_{22} = \varepsilon_1 + \chi_e\left(\frac{q^2}{k^2}\right)^{s-1}, \quad \varepsilon_{12} = i\varepsilon_2 - i\chi_e\left(\frac{q^2}{k^2}\right)^{s-1}, \quad (4.20)$$

where $\chi_e = -\frac{N_e}{2s!h_e}\left(\frac{s^2\beta_e^2}{2}\right)^{s-1}$, $N_e = \frac{\Omega_e^2}{\omega_e^2}$, $\beta_e = \frac{v_{Te}}{c}$, and $\varepsilon_1$ and $\varepsilon_2$ are the components of the cold magnetoactive plasma permittivity tensor [24–27, 35].

The dispersion relation for electron SCX-modes in the structure plasma–metal is derived as the result of applying the following boundary condition. The tangential electric field of the SCX-mode should be equal to zero at the metal interface due to the high electric conductivity of the metal. The dispersion relation has the following form:

$$\frac{k_2^2}{2i|k_2|k\varepsilon_{11}(i|k_2|)} = \sum_{j=0}^{s-2}\frac{k_2^2q_j}{2(s-1)\varepsilon_1k\left(k_2^2 + q_j^2\right)}. \quad (4.21)$$

As it follows from the analysis of (4.21), this equation has solutions only for even numbers $s$ of electron cyclotron frequency harmonics and negative values of the wavenumber $k_2$. In other words SCX-modes are unidirectional waves which propagate in the direction opposite to that of the $\vec{y}$ axis (wavenumber $k_2 < 0$). Their frequency increases with increasing absolute value of the wavenumber $|k_2|$. Simple solutions of the dispersion relation (4.21) can be obtained in the case of weak spatial dispersion of the plasma along the plasma interface when the following inequality is valid: $|k_2\rho_e| \ll 1$. These solutions are as follows:

$$h_e \approx \begin{cases} 0.5 \cdot k_2^2 \rho_e^2, & s = 2, \\ 0.005 \cdot k_2^6 \rho_e^6, & s = 4. \end{cases} \tag{4.22}$$

From the analysis of (4.22) it follows, that SCX-modes are characterized by normal dispersion, i.e. the directions of the wave phase and group velocity coincide. These wave frequencies are a little bit higher than the corresponding cyclotron harmonic frequency $s|\omega_e|$. The depth of their penetration into the plasma is of the order of the wavelength $\left(\left|q_j\right| \approx |k_2|\right)$ which is much larger than the electron Larmor radius. Unlike in the case of the plasma–dielectric structure, for which SCX-modes propagate in both directions along the plasma interface (see Sect. 4.1), in the waveguide structure plasma–metal, SCX-modes are unidirectional (they propagate in one direction only).

The damping of these waves is caused by three reasons: spatial dispersion of the medium, presence of particle collisions and dissipation of the wave energy in the metal (due to the finite metal electrical conductivity). To calculate the SCX-mode damping rate $\delta_{\text{m}}$ caused by the energy dissipation in the metal one has to apply the Fermi distribution function [27] for describing the conductivity of the electrons in the metal. Then via introducing the effective frequency $\nu_{\text{m}}$ of electron collisions in the metal, one can estimate the value $\delta_{\text{m}}$ as follows: $\delta_{\text{m}} \sim \nu_{\text{m}} n_{\text{pl}}/n_{\text{m}}$, where $n_{\text{pl}}$ and $n_{\text{m}}$ are the electron particle density in plasma and metal, respectively [32, 36]. This does not need much attention since it is clear that the value $\delta_{\text{m}}$ is very small due to presence of the factor $n_{\text{pl}}/n_{\text{m}}$. For typical magnitudes of electron particle densities in plasma and metal, this factor is of the order of $10^{-10}$ to $10^{-12}$.

To calculate the SCX-mode damping rate $\delta_{\nu e}$ caused by plasma particle collisions, one has to account for the collisional frequency $\nu_e$ [26, 27] in the tensor $\varepsilon_{ik}$ components. Then via replacing $k_2$ by the sum $k_2 + i\delta_{\nu e}$ ($|\delta_{\nu e}| \ll |k_2|$) one can derive $\delta_{\nu e}$ from the dispersion relation (4.21):

$$\delta_{\nu e} = \frac{\nu_e |k_2|}{2(s-1)\omega h_e}. \tag{4.23}$$

The weak spatial plasma dispersion results in the third term $\delta_{pe}$ in the total value of the SCX-mode damping rate whose magnitude is found to be also small. This partial damping rate $\delta_{pe}$ in the case of $s = 2$ can be calculated, keeping in mind that the numerical coefficients in the expressions for the SCX-mode damping rate are determined by the value $Q_{ik}$ and decrease by an order of magnitude, when the electron cyclotron frequency increases by a factor of two:

$$\delta_{pe} \approx 1.67 \cdot \sqrt{h_e/(3\pi)}|k_2|, \quad k_2 < 0. \tag{4.24}$$

As it follows from analysis of (4.24), no significant difference is introduced in the value of this SCX-mode damping rate when one replaces the plasma–dielectric waveguide structure [33, 37] by the structure plasma–metal.

Thus, surface eigenwaves of electron cyclotron type with extraordinary polarization are shown to propagate in the waveguide structures uniform plasma–metal under the conditions of an external stationary magnetic field parallel with respect to the plasma interface. They are weakly damped and their dispersion is normal. It should be underlined that these waves propagate across $\vec{B}_0$ in one direction only. This direction coincides with that of electron Larmor gyration nearby the plasma interface. That is why these waves perfectly transfer the energy in one direction, and there is no backward wave for them as a consequence of this property of unidirectionality.

## 4.3 Influence of Finite Waveguide Transverse Dimensions and Inhomogeneity of Plasma Particle Density on Dispersion Properties of Surface Electron Cyclotron X-Modes

In this section, propagation of electron SCX-mode waves in nonuniform waveguide structures plasma–dielectric–metal is analyzed. Oxide films on the waveguide metal wall resulting from plasma-wall interaction or coating of the metal surface with a protective dielectric layer significantly change the dispersion properties of surface waves propagating in such waveguides as compared to the case of waveguide structures with direct contact of the plasma with the metal wall [5, 13–15, 33, 38]. In plasma–dielectric–metal structures, there is the additional possibility for SCX-modes to propagate in two opposite directions across $\vec{B}_0$. This phenomenon is closely connected with the presence of the external stationary magnetic field $\vec{B}_0$, which is oriented along the plasma interface (Voigt geometry). The frequencies of the waves, which propagate in mutually opposite directions with the same absolute value of the wavenumber, are different. Electron SCX-modes at the interface semibounded plasma–dielectric appear to be non-mutual ones [33]. This property is also inherent to SCX-modes in the case of the three-component waveguide structure plasma–dielectric–metal [39].

Here, a semibounded plasma, which occupies the half-space $x > 0$, is considered (see Fig. 4.3). The $x = 0$ plane is the interface between the plasma and the dielectric layer with thickness $a_\text{d}$ and dielectric constant $\varepsilon_\text{d}$. The interface between the dielectric and the metal wall is the plane $x = -a_\text{d}$. The external stationary magnetic field $\vec{B}_0$ is oriented along the $\vec{z}$ axis. The plasma is assumed to be uniform along the $z$ direction and non-uniform along the $x$ axis. The plasma spatial dispersion is assumed to be weak. Then one can get use of the expressions (4.20) for the plasma permittivity tensor $\varepsilon_{ik}$, which are derived in Sect. 4.2.

To get the dispersion relation for SCX-modes which propagate in such structures one has to solve (4.1) and get the expressions for the tangential field components of SCX-modes in the dielectric region $-a_d < x < 0$. For this purpose one can apply (4.4) with the following replacements: $\varepsilon_{11} \to \varepsilon_{\mathrm{d}}$, $\varepsilon_{33} \to \varepsilon_{\mathrm{d}}$, $\varepsilon_{13} \to 0$. Then one has to set the plasma impedance equal to that of the dielectric at the boundary of these two media (plane $x = 0$) and set the tangential component of the SCX-mode electric field to zero at the interface dielectric–metal (plane $x = -a_{\mathrm{d}}$). This corresponds to application of the following two boundary conditions, which are typical for problems of linear plasma electrodynamics.

- The tangential components of SCX-mode electric and magnetic fields should be continuous at the plasma–dielectric interface.
- The tangential component of the wave electric field should vanish (be equal to zero) at the dielectric–metal interface since the electric conductivity of the metal is assumed to be infinitely high.

Solving (4.1) in the region of dielectric, $-a_{\mathrm{d}} < x < 0$, makes it possible to derive the surface impedance $Z_{\mathrm{d}}(0) = E_y(0)/H_z(0)$ of the dielectric layer:

$$Z_{\mathrm{d}}(0) = -i|k_2|(k\varepsilon_{\mathrm{d}})^{-1} \tanh(|k_2|a_{\mathrm{d}}). \tag{4.25}$$

The surface impedance $Z_{\mathrm{pl}}(0) = E_x(0)/H_y(0)$ of the semibounded plasma can be obtained from solving (4.1) in the plasma region, $x > 0$:

$$Z_{\mathrm{pl}}(0) = i|k_2|\left(k\sqrt{\varepsilon_{11}(\varepsilon_{33} + 2B)}\right)^{-1}, \tag{4.26}$$

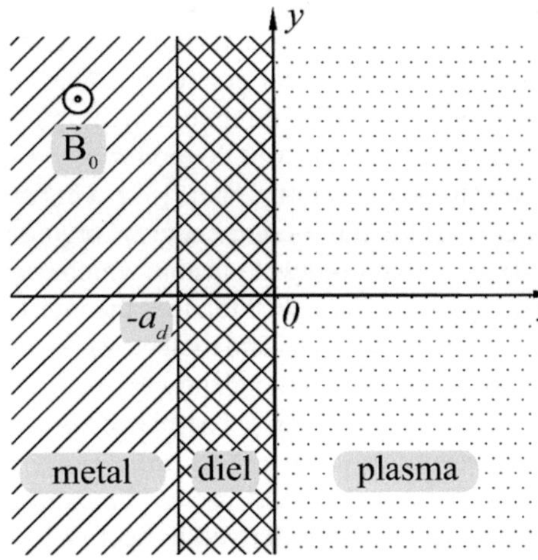


**Fig. 4.3** Schematic of the problem "semibounded plasma–dielectric–metal"

where $B = \varepsilon_{13}k_2/k_3$ is a value which does not depend on the wavenumber of the SCX-mode. Using (4.25) and (4.26), one gets the dispersion relation for SCX-modes in this waveguide structure:

$$\sqrt{\varepsilon_{11}(\varepsilon_{33}+2B)} = \varepsilon_{\mathrm{d}} \coth(|k_2|a_{\mathrm{d}}). \tag{4.27}$$

There are two ways for taking into account plasma particle density inhomogeneity. The first one consists in solving the kinetic equation for plasma particles for arbitrary plasma particle density. The second way is to model the density inhomogeneity by a multicomponent structure in which each layer is characterized by its own width and particle density. Mathematical simplicity dictates to use the second way. Nevertheless, it is cumbersome to obtain the dispersion relations. The latter is associated with multiple application of the boundary conditions for the wave fields.

To deal with less cumbersome expressions one should just take the simplest case of an inhomogeneous plasma which consists of one transient layer of the thickness $a_1$ with particle density $n_1 \neq n_0$, where $n_0$ is the particle density in the region $x \geq a_1$. The SCX-mode electromagnetic field is described by the Maxwell equations. In the approach of slow waves where the phase velocity satisfies the inequality $\upsilon_{\mathrm{ph}}^2|\varepsilon_{ik}| \ll c^2$, the following expression for the impedance $Z_{\mathrm{pl0}}$ of the semibounded plasma at the interface $x = a_1$ can be derived with the aid of Fourier analysis:

$$Z_{\mathrm{pl0}}(a_1) = \frac{E_y(a_1)}{H_z(a_1)} = \frac{i[(s-1)-R]k_2}{[(\varepsilon_1+\varepsilon_2)R+(s-1)(\varepsilon_2+\eta\varepsilon_1)]k}, \tag{4.28}$$

where $R = \sum_{j=0}^{s-2} \frac{i|k_2|q_j}{q_j^2+k_2^2}$, $q_j = \frac{1}{\rho_e}\left[\frac{|\varepsilon_1 h_e| 2^{2s-1} s! N_e^{-2}}{\exp(-2\pi i(2j+1))}\right]^{\frac{1}{2(s-1)}}$, $\eta = \frac{|k_2|}{k_2}$, and $\varepsilon_1$ and $\varepsilon_2$ are the components of the magnetoactive plasma permittivity tensor in hydrodynamic approach [25–27, 34, 35].

The impedance $Z_{\mathrm{d}}(0)$ of the dielectric at the interface $x = 0$ is already known in the approach of slow waves [see (4.25)].

One can derive the expressions for the impedances of the transient layer at the planes $x = 0$ and $x = a_1$ on the base of the expression for $E_y(x)$:

$$E_y(x) = \sum_{n=-\infty}^{\infty} \exp(i\alpha_n x)\Big[\frac{ik_2^2}{k\Delta(\alpha_n)}\big(H_z(0)-(-1)^n H_z(a_1)\big) + \Delta^{-1}(\alpha_n) i (k_2\varepsilon_{12} - \alpha_n\varepsilon_{11})\big(E_y(0)-(-1)^n E_y(a_1)\big)\Big], \tag{4.29}$$

where $\Delta(\alpha_n) = \varepsilon_{11}(\alpha_n)\left(\alpha_n^2 + k_2^2\right)$. To derive the dispersion relation of SCX-modes one has to set equal the impedances of the transient layer and the corresponding impedances which are calculated for the planes $x = 0$ and $x = a_1$.

Since the dispersion relation has a very cumbersome form for arbitrary cyclotron harmonic $s$, and the relation cannot be solved analytically in this case, only the case of $s = 2$ is considered here. However, this does not mean that the problem cannot be solved for some higher numbers of cyclotron harmonic. From practical point of view, the case of $s = 2$ is the most interesting one. For instance, additional plasma heating in fusion devices is most often carried out under the conditions of electron cyclotron resonance at just the second harmonic [40]. Moreover, damping of this harmonic caused by the collisions of plasma particles and their interaction with the plasma interface is lower compared to the higher harmonics (for fixed values of wavelength and other parameters of the waveguide structure) [32, 33, 39].

In this case the dispersion relation for SCX-modes can be written as:

$$
\begin{aligned}
&\frac{\tanh(|k_2|a_\mathrm{d})}{\varepsilon_\mathrm{d}}F\left\{1+\left[\frac{\eta\varepsilon_2-\varepsilon_1}{2\varepsilon_1\sinh(|k_2|a_1)}+\frac{q}{1-q_1^2}\frac{\eta(\varepsilon_2+\varepsilon_1)}{2\varepsilon_1\sinh(|k_2|q_1a_1)}\right]\right.\\
&\quad\left.-\left[\frac{\eta\varepsilon_2-\varepsilon_1}{2\varepsilon_1\tanh\,(|k_2|a_1)}+\frac{q_1}{1-q_1^2}\frac{\eta(\varepsilon_2+\varepsilon_1)}{2\varepsilon_1\tanh(|k_2|q_1a_1)}\right]^2\right\}\\
&\quad+\left[\frac{\coth(|k_2|a_1)}{2\varepsilon_1}+\frac{q_1}{1-q_1^2}\frac{\coth(|k_2|q_1a_1)}{2\varepsilon_1}\right]^2\\
&\quad-\left[\frac{1}{2\varepsilon_1\sinh(|k_2|a_1)}+\frac{q_1}{1-q_1^2}\frac{1}{\sinh(|k_2|q_1a_1)}\right]^2+\left(F+\frac{\tanh(|k_2|q_1a_1)}{2\varepsilon_1}\right)\\
&\quad\times\left[\frac{\coth(|k_2|a_1)}{2\varepsilon_1}+\frac{q_1}{1-q_1^2}\frac{\coth(|k_2|q_1a_1)}{2\varepsilon_1}\right]-\left(F-\frac{\tanh(|k_2|a_\mathrm{d})}{\varepsilon_\mathrm{d}}\right)\\
&\quad\times\left\{\left[\frac{\coth(|k_2|a_1)}{2\varepsilon_1}+\frac{q_1}{1-q_1^2}\frac{\coth(|k_2|q_1a_1)}{2\varepsilon_1}\right]\right.\\
&\quad-\left[\frac{\eta\varepsilon_2-\varepsilon_1}{2\varepsilon_1\tanh(|k_2|a_1)}+\frac{q_1}{1-q_1^2}\frac{\eta(\varepsilon_2+\varepsilon_1)}{2\varepsilon_1\tanh(|k_2|q_1a_1)}\right]\\
&\quad-\left[\frac{1}{2\varepsilon_1\sinh(|k_2|a_1)}+\frac{q_1}{1-q_1^2}\frac{1}{2\,\varepsilon_1\sinh(|k_2|q_1a_1)}\right]\\
&\quad\left.\times\left[\frac{\eta\varepsilon_2-\varepsilon_1}{2\varepsilon_1\sinh(|k_2|a_1)}+\frac{q_1}{1-q_1^2}\frac{\eta(\varepsilon_2+\varepsilon_1)}{2\varepsilon_1\sinh(|k_2|q_1a_1)}\right]\right\}=0. \qquad (4.30)
\end{aligned}
$$

where $q_i = q(n_i) = \frac{4}{|k_2|}\sqrt{-\frac{\varepsilon_1(n_i)h_e}{N_e^2(n_i)}}$, and $F(n_1) = \frac{1+q_2/(1-q_2^2)}{\varepsilon_1+\eta\varepsilon_2+\eta(\varepsilon_2+\varepsilon_1)\left[q_2/(1-q_2^2)\right]}$. This (4.30) can be solved numerically only. Doing that, it is convenient to describe the properties of the dielectric by the following parameter: $\chi_\mathrm{d} = \varepsilon_\mathrm{d}\rho_e a_\mathrm{d}^{-1}$, and the plasma inhomogeneity by the dimensionless parameter $\zeta_p = \left(n_0 n_1^{-1} - 1\right)\rho_e a_1^{-1}$. The results of this study are presented in Figs. 4.4, 4.5 and 4.6 for $N_e^2 = 10^3$ and $\chi_\mathrm{d} = 1$. As it follows from the figures, five dispersion curves correspond to the same set of waveguide parameters. These five solutions of the dispersion relation (4.30) correspond to the waves which propagate in the transient layer (along its two interfaces) and in the region of semi-bounded plasma. However, only one of them, which is marked by the symbol "*", can be called the main mode. It propagates along the interface dielectric–metal, and its properties depend mostly on the dielectric characteristics. The other four solutions correspond to additional modes whose existence is caused by the choice of the model for the plasma density inhomogeneity.

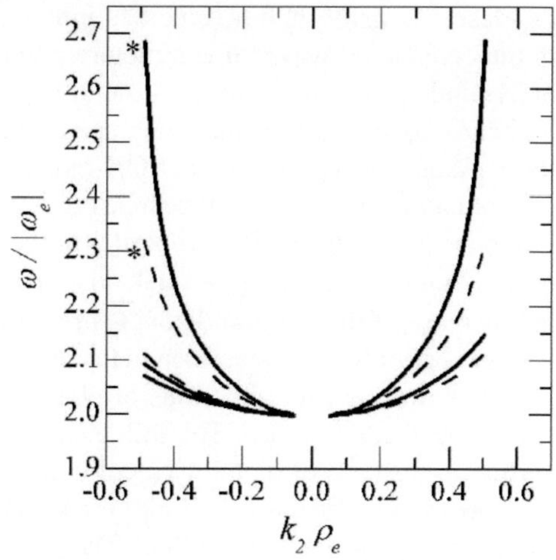


**Fig. 4.4** SCX-mode frequency versus wave number. Solid lines correspond to the case $\zeta p = 0.2$, dashed lines to $\zeta p = 20$

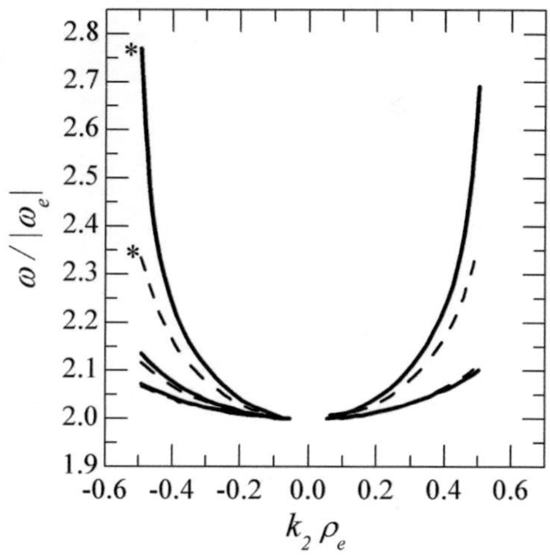


**Fig. 4.5** SCX-mode frequency versus wave number. Solid lines correspond to the case $\zeta p = -0.07$ and dashed lines to $\zeta p = -7$

Thus introducing the dielectric layer between the region of plasma and the waveguide metal wall significantly changes the SCX-mode dispersion properties as compared to the case studied in Sect. 4.2. First of all, new branches of SCX-modes appear which propagate with positive values of the wavenumber $k_2$. These are two additional modes, whereas three modes (including the main one) can propagate with negative values of the wavenumber $k_2$. For fixed values of wavelengths, SCX-modes which propagate against the $\vec{y}$ axis (i.e. with wavenumbers $k_2 < 0$) have eigenfrequencies which are different from those of SCX-modes which propagate with $k_2 > 0$ (along the $\vec{y}$ axis). If the thickness of the dielectric layer is gradually decreased, the wave with the shortest wavelength remains in the structure plasma–metal. Decreasing dielectric layer thickness is equivalent to increasing dielectric constant $\varepsilon_d$. A change in the parameter $\chi_d$ results in a change of the phase velocities of the additional modes only.

Plasma density inhomogeneities influence the SCX-mode dispersion properties in the considered waveguide structure practically in the same way as in the case of SCX-mode propagation in the structures plasma–metal. First of all, new dispersion branches appear in the frequency spectrum of SCX-modes. The phase velocity of these modes decreases and the SCX-mode damping rate caused by the plasma particle collisions increases with decreasing averaged plasma particle density. The main influence on the dispersion properties of these waves is caused by $\vec{B}_0$. The dispersion of SCX-modes is normal for relatively weak magnetic fields, when $\Omega_e^2 \gg \omega_e^2$. Gradual increase of the magnitude of $\vec{B}_0$ first results in increasing SCX-mode frequency, and then, after linear conversion to the upper hybrid mode, the SCX-mode dispersion becomes reverse (the directions of phase and group velocities become opposite to each other), see Fig. 4.6. The influence of plasma density inhomogeneity is weak, similar to the case of metal waveguide without dielectric coating.

Now the influence of the finite transverse dimension of the waveguide on SCX-mode propagation will be studied. The plasma filling of the waveguide is assumed to be uniform. The coordinate system is chosen such that the plane plasma occupies the

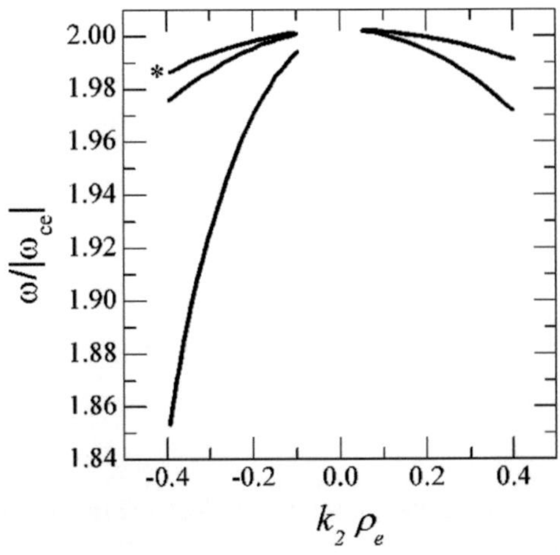


**Fig. 4.6** SCX-mode frequency versus wave number; $N_e^2 = 0.5$, $\chi_d = 1$ and $\zeta p = 0.2$

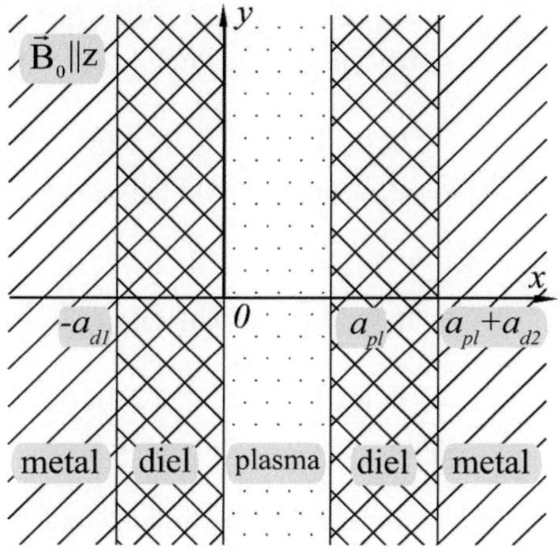


**Fig. 4.7** Schematic of the problem "influence of finite transverse dimension of waveguide on SCX-mode propagation"

region $0 \leq x \leq a_{\mathrm{pl}}$ (see Fig. 4.7). The dielectric layers have the thicknesses $a_{\mathrm{d1}}$ and $a_{\mathrm{d2}}$. Then the coordinates of the internal surfaces of the metal walls are $x = -a_{\mathrm{d1}}$ and $x = a_{\mathrm{pl}} + a_{\mathrm{d2}}$. The Fourier coefficient of the SCX-mode tangential electric field in the plasma layer can be written in the following form:

$$E_2 = \frac{[(-1)^n E_y(a_{\mathrm{pl}}) - E_y(0)][k_2\varepsilon_{12}(\alpha_n) - \alpha_n\varepsilon_{11}(\alpha_n)]}{iak^2\Delta(\alpha_n)} + \frac{[(-1)^n H_z(a_{\mathrm{pl}}) - H_z(0)]k_2^2}{iak^3\Delta(\alpha_n)}. \tag{4.31}$$

The notations applied her are analogous to those applied earlier in this section.

After carrying out the inverse Fourier transform of (4.31) one gets the set of two equations for the SCX-mode electric field $E_y$ at the plasma layer interfaces $x = 0$ and $x = a_{\mathrm{pl}}$. These equations include four unknown values of the waves $E_y$ and $H_z$ at the two interfaces:

$$\begin{aligned} 2\varepsilon_1 E_y(0) &= \frac{|k_2|}{ik}B_1 - (\varepsilon_1 - \eta\varepsilon_2)A_1 - \frac{i|k_2|}{k}\frac{q}{1-q^2}B_2 + \eta(\varepsilon_1+\varepsilon_2)\frac{q}{1-q^2}A_2, \\ 2\varepsilon_1 E_y(a_{\mathrm{pl}}) &= \frac{|k_2|}{ik}B_3 - (\varepsilon_1 - \eta\varepsilon_2)A_3 - \frac{i|k_2|}{k}\frac{q}{1-q^2}B_4 + \eta(\varepsilon_1+\varepsilon_2)\frac{q}{1-q^2}A_4, \end{aligned} \tag{4.32}$$

where

$$A_1 = \frac{E_y(a_{\rm pl})}{\sinh(|k_2|a_{\rm pl})} - \frac{E_y(0)}{\tanh(|k_2|a_{\rm pl})}, \quad B_1 = \frac{H_y(a_{\rm pl})}{\sinh(|k_2|a_{\rm pl})} - \frac{H_y(0)}{\tanh(|k_2|a_{\rm pl})},$$

$$A_2 = \frac{E_y(a_{\rm pl})}{\sinh(q|k_2|a_{\rm pl})} - \frac{E_y(0)}{\tanh(q|k_2|a_{\rm pl})}, \quad B_2 = \frac{H_y(a_{\rm pl})}{\sinh(q|k_2|a_{\rm pl})} - \frac{H_y(0)}{\tanh(q|k_2|a_{\rm pl})},$$

$$A_3 = \frac{E_y(a_{\rm pl})}{\tanh(|k_2|a_{\rm pl})} - \frac{E_y(0)}{\sinh(|k_2|a_{\rm pl})}, \quad B_3 = \frac{H_y(a_{\rm pl})}{\tanh(|k_2|a_{\rm pl})} - \frac{H_y(0)}{\sinh(|k_2|a_{\rm pl})},$$

$$A_4 = \frac{E_y(a_{\rm pl})}{\tanh(q|k_2|a_{\rm pl})} - \frac{E_y(0)}{\sinh(q|k_2|a_{\rm pl})}, \quad B_4 = \frac{H_y(a_{\rm pl})}{\tanh(q|k_2|a_{\rm pl})} - \frac{H_y(0)}{\sinh(q|k_2|a_{\rm pl})}.$$

The solutions of the Maxwell equations for the SCX-mode fields in the dielectric layers make it possible to find the surface impedances at the planes $x = 0$ and $x = a_{\rm pl}$. This allows to exclude two unknown values from the set of (4.32). The expressions for $Z_{\rm d}(0)$ and $Z_{\rm d}(a_{\rm pl})$ are:

$$\begin{cases} Z_{\rm d}(0) = |k_2| \tanh(|k_2|a_{\rm d_1})/(ik\varepsilon_{\rm d_1}), \\ Z_{\rm d}(a_{\rm pl}) = -|k_2| \tanh(|k_2|a_{\rm d_2})/(ik\varepsilon_{\rm d_2}). \end{cases} \tag{4.33}$$

To derive the expressions (4.33) one has to apply the boundary conditions for the tangential components of the SCX-mode electric field at the metal interfaces: $E_y(-a_{\rm d_1}) = E_y(a_{\rm pl} + a_{\rm d_2}) = 0$. After solving the set of equations (4.32) taking (4.33) into account, makes it possible to derive the SCX-mode dispersion relation in the considered waveguide structure:

$$\left(T_1 + \frac{T\varepsilon_{\rm d_1}}{\tanh(|k_2|a_{\rm d_1})} + 2\varepsilon_1\right)\left(T_1 - \frac{T\varepsilon_{\rm d_2}}{\tanh(|k_2|a_{\rm d_2})} + 2\varepsilon_1\right)$$
$$= \left(U + \frac{\varepsilon_{\rm d_1}}{\tanh(|k_2|a_{\rm d_1})}\right)\left(U_1 - \frac{\varepsilon_{\rm d2}}{\tanh(|k_2|a_{\rm d_2})}\right), \tag{4.34}$$

where the following notations are applied:

$$U = \frac{1}{\sinh(|k_2|a_{\rm pl})} + \frac{q}{(1-q^2)\sinh(|k_2|qa_{\rm pl})},$$
$$U_1 = \frac{\eta\varepsilon_2 - \varepsilon_1}{\sinh(|k_2|a_{\rm pl})} + \frac{\eta(\varepsilon_2 + \varepsilon_1)q}{(1-q^2)\sinh(|k_2|qa_{\rm pl})},$$
$$T = \frac{1}{\tanh(|k_2|a_{\rm pl})} + \frac{q}{(1-q^2)\tanh(|k_2|qa_{\rm pl})},$$
$$T_1 = \frac{\eta\varepsilon_2 - \varepsilon_1}{\tanh(|k_2|a_{\rm pl})} + \frac{q\eta(\varepsilon_2 + \varepsilon_1)}{(1-q^2)\tanh(|k_2|qa_{\rm pl})}.$$

As it follows from analysis of (4.34), the relation has solutions for both $k_2 > 0$ and $k_2 < 0$ unlike in the case of SCX-mode propagation in metal waveguides completely filled by plasma. In this case SCX-modes are non-mutual which means that the dispersion properties of the waves with the same absolute values and opposite signs of the wavenumbers are different. The presence of dielectric layers influences the shape of the dispersion curves weakly. This is explained by the fact that the SCX-mode field sharply decreases with going away from the boundaries of the media. Note that in the limit of zero thickness of the dielectric layer only one solution of the two possible with $k_2 > 0$ and $k_2 < 0$ remains, that with the shortest wavelength and $k_2 < 0$.

Thus, unlike metal waveguides which are entirely filled by plasmas, in the waveguide structures plasma–dielectric–metal, SCX-modes can propagate across the external stationary magnetic field with both signs of the wavenumber: $k_2 > 0$ and $k_2 < 0$. The frequency of the SCX-modes with negative wavenumber ($k_2 < 0$) is different from that of the SCX-modes with positive wavenumber of the same absolute value, i.e. no degeneracy of these modes is observed. Making the dielectric layer thinner (increasing the parameter $\chi_{\mathrm{d}} = \varepsilon_{\mathrm{d}} \rho_e a_{\mathrm{d}}^{-1}$) results in vanishing of the mode with positive wavenumber ($k_2 > 0$).

The depths of SCX-mode penetration into both plasma and dielectric are of the order of the wavelengths. A decrease of the thickness of dielectric coating at the metal interface leads to vanishing of the SCX-mode with the longest wavelength (among those existing). The remaining mode propagates in the same direction in which plasma electrons gyrate along the Larmor orbits nearby the waveguide metal wall. The influence of the transverse dimensions of the plasma layer on the SCX-mode dispersion properties can be neglected until the plasma thickness becomes of the order of the SCX-mode penetration depth into the plasma. Then the degeneration of these waves frequency removes.

## 4.4 Parametric Excitation of Surface Electron Cyclotron X-Modes

The parametric effect of an external RF field on plasmas has been studied for a long time [41, 42]. In this frame, the case of bulk electron cyclotron wave excitation was investigated quite well, which is explained by the fact, that these waves are practically applied for plasma heating, in particular, in controlled thermonuclear fusion [40, 43–55]. Excitation of SWs has its peculiarities as compared to the case of the bulk wave excitation [56–61]. The growth rates of parametric instabilities have different functional dependences on the pumping wave amplitude in these cases. Growth rates of SWs essentially depend on parameters of the waveguide and plasma filling. So this section is devoted to studying the parametric excitation of electron SCX-modes at the harmonics of the electron cyclotron frequency in planar dielectric waveguide with uniform plasma filling. The set of equations which describe the initial stage of the studied wave parametric instability is derived and analyzed both analytically (in the limiting cases of thick and thin plasma layer) and numerically.

The considered uniform plasma layer occupies the space $0 < x < a_{\rm pl}$ (see Fig. 4.8). At the planes $x = 0$ and $x = a_{\rm pl}$ there are interfaces to layers of a dielectric medium with dielectric constant $\varepsilon_{\rm d}$. The external stationary magnetic field $\vec{B}_0$ is oriented along the $\vec{z}$ axis. A monochromatic electric pumping field $\vec{E}_0 \cos(\omega_0 t)$ (here $\vec{E}_0$ is the amplitude, and $\omega_0$ is the frequency) is present at the interface $x = a_{\rm pl}$. The field has no component along the external magnetic field. The surface modes under study in this section have the field components $\left(E_x, E_y, H_z\right)$. They propagate in the direction of the $\vec{y}$ axis across the magnetic field $\vec{B}_0$. The dependence of the SCX-mode fields on coordinates and time is assumed in the following form:

$$E, H = \sum_{n=-\infty}^{+\infty} E^{(n)}, H^{(n)} \exp(ik_2 y - i(\omega + n\omega_0)t). \tag{4.35}$$

To describe these wave fields, one uses the set of Maxwell equations, which have to be solved employing the Fourier method. The following set of equations can be derived for the Fourier coefficients of the SCX-mode fields in the region occupied by the plasma:

$$\begin{cases} ia\alpha_n H_3^{(p)} = H_z^{(p)}(0) - (-1)^n H_z^{(p)}\left(a_{\rm pl}\right) + ika_{\rm pl} E_1^{(p)} - (4\pi/c)a_{\rm pl} j_2^{(p)}, \\ k_2 H_3^{(p)} = -kE_1^{(p)} - (4\pi i/c) j_1^{(p)}, \\ -ika_{\rm pl} H_3^{(p)} = E_y^{(p)}(0) - (-1)^n E_y^{(p)}\left(a_{\rm pl}\right) - ia_{\rm pl}\alpha_n E_2^{(p)} + ia_{\rm pl} k_2 E_1^{(p)}, \end{cases} \tag{4.36}$$

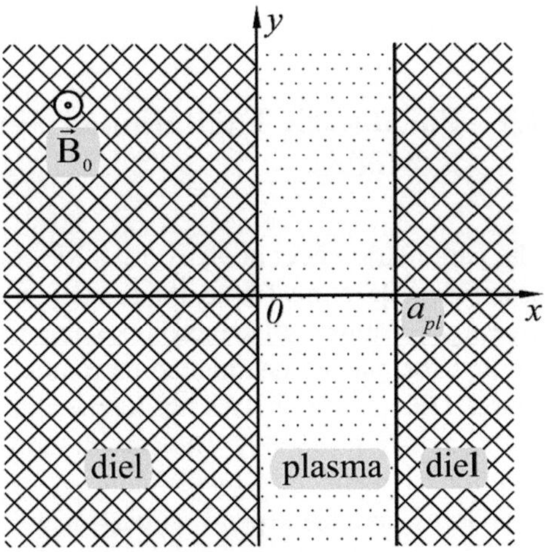


**Fig. 4.8** Schematic of the problem "parametric excitation of SCX-modes"

where $\alpha_n = \pi n a_{\rm pl}^{-1}$, $n = 0, \pm 1, \pm 2, \ldots$, $E_y(x = 0, x = a_{\rm pl})$ and $H_z(x = 0, x = a_{\rm pl})$ are SCX-mode fields at the interfaces of the plasma layer, and $j_1^{(p)}$ and $j_2^{(p)}$ are the Fourier coefficients of the electric current density.

To determine $j_1^{(p)}$ and $j_2^{(p)}$ one has to solve the kinetic Vlasov–Boltzmann equation for the plasma particle distribution function $f_\alpha$ with account for the external RF pumping field:

$$\frac{\partial f_\alpha}{\partial t} + \vec{\upsilon}\frac{\partial f_\alpha}{\partial \vec{r}} + \frac{e_\alpha}{m_\alpha}\left\{\vec{E}_0\cos(\omega_0 t) + \frac{1}{c}\left[\vec{\upsilon}, \vec{B}_0\right]\right\}\frac{\partial f_\alpha}{\partial \vec{\upsilon}}$$
$$= -\frac{e_\alpha}{m_\alpha}\left\{\vec{E} + \frac{1}{c}\left[\vec{\upsilon}, \vec{B}_0\right]\right\}\frac{\partial f_{\alpha 0}}{\partial t}. \tag{4.37}$$

The unperturbed distribution function $f_{\alpha 0}(\vec{\upsilon}, t)$ can be expressed in terms of a Maxwellian one $f_\alpha^{\rm M}$:

$$f_{\alpha 0}(\vec{\upsilon}, t) = f_\alpha^{\rm M}\left(\vec{\upsilon} + \frac{e_\alpha \vec{E}_0}{m_\alpha \omega_0}\cos(\omega_0 t)\right). \tag{4.38}$$

In the approximation of weak plasma spatial dispersion one can obtain the following expressions for the Fourier coefficients of the electric current density $\vec{j} = \sum\limits_{p=-\infty}^{+\infty} \vec{j}^{(p)} \exp(-i\omega t + ip\omega_0 t)$:

$$\begin{cases} j_1^{(p)} = \sigma_1(s, m, l, p)E_1^{(p+l)} + \sigma_2(s, m, l, p)E_2^{(p+l)}, \\ j_2^{(p)} = \sigma_1(s, m, l, p)E_2^{(p+l)} - \sigma_2(s, m, l, p)E_1^{(p+l)}, \end{cases} \tag{4.39}$$

where

$$\sigma_1(s, m, l, p) = \sum_\alpha \frac{\Omega_\alpha^2}{4\pi} \sum_{s,m,l=-\infty}^{+\infty} \frac{is^2 e^{-y_\alpha} I_s(y_\alpha)}{\left(\omega_{p+m} - s\omega_\alpha\right) y_\alpha} e^{-i\phi l} J_m(g) J_{m-1}(g),$$

$$\sigma_2(s, m, l, p) = -\sum_\alpha \frac{\Omega_\alpha^2}{4\pi} \sum_{s,m,l=-\infty}^{+\infty} \frac{s e^{-y_\alpha}\left(I_s'(y_\alpha) - I_s(y_\alpha)\right)}{\omega_{p+m} - s\omega_\alpha} e^{-i\phi l} J_m(g) J_{m-1}(g),$$

$$\phi = \operatorname{arc} tg\frac{E_{0y}}{E_{0x}}, \; g = k_2\beta_E \equiv k_2\rho_e g_0 = \sqrt{\frac{k_2^2 e_\alpha^2\left(\omega_0^2 E_{0x}^2 + \omega_\alpha^2 E_{0y}^2\right)}{m_\alpha^2 \omega_0^2\left(\omega_0^2 - \omega_\alpha^2\right)^2}}, \; y_\alpha = \frac{\alpha_n^2 \rho_\alpha^2}{2} \ll 1,$$

$\omega_{p+m} = \omega + (p + m)\omega_0$, and $J_m(x)$ are Bessel functions of the first kind. The summation over the indices *s, m,* and *l* should be done independently in the limits from $-\infty$ to $+\infty$.

To get an infinite set of equations for the $E_y$-field component of SCX-modes one has to apply the following boundary conditions. The tangential component of the

SCX-mode electric field should be continuous at the interface of plasma layer and dielectric medium. The tangential component of the SCX-mode magnetic field has a discontinuity at the plane $x = a_{\text{pl}}$, which is caused by the presence of a nonlinear electric surface current at this interface as the result of the action of the external RF pumping field. The field $H_z(0)$ in the set of equations (4.36) can be excluded after solving the Maxwell equations in the region of the dielectric ($x \le 0$). Then the following formula can be derived:

$$H_z^n(0) = \frac{i(\omega + n\omega_0)}{|k_2|c} E_y^n(0). \tag{4.40}$$

After integrating (4.36) using relation (4.40), one can find the expression for the second boundary condition mentioned above:

$$H_z^n(a_{\text{pl}}) = -\frac{i\omega_n}{|k_2|c} E_y^n(a_{\text{pl}}) + \frac{N_e^2}{c} \sum_{m,l} \frac{a_E^{|m|+|m-l|} e^{-i\phi l}}{m!(m-l)!k_2} i^{|m|+|m-l|+1}$$
$$\times \left( \frac{\omega_e^3}{\omega_{n+m}^2 - \omega_e^2} + \frac{k_2^2 \rho_e^2 \omega_e^2}{4(\omega_{n+m} - 2|\omega_e|)} \right) E_y^{n+l}(a_{\text{pl}}), \tag{4.41}$$

where $a_E = 0.5 b_E |k_2| \rho_e$.

When deriving (4.41) the assumption of small amplitude of the pumping RF field is applied. This is justified by the fact that the amplitude of plasma electron oscillations in the external pumping RF field is smaller or of the order of their Larmor radius of gyration in the external stationary magnetic field.

After application of the inverse Fourier transform to (4.36) with account for (4.40) and (4.41) one can derive the following infinite set of equations which describe parametric excitation of SCX-modes at the second harmonic of the electron cyclotron frequency:

$$D^{p+l} E_y^{p+l}(a_{\text{pl}}) - F_{m,l} E_y^{p+l} = 0, \tag{4.42}$$

where

$$G_{m,l} = \sum_{m,l} \frac{a_E^{|m|+|m-l|} e^{-i\phi l}}{m!(m-l)!} \left( \frac{k_2 \Omega_e^2 \omega_e \omega_p^{-1}}{\omega_{p+m}^2 - \omega_e^2} + \frac{\Omega_e^2 \omega_p^{-1} k_2^3 \rho_e^2}{4(\omega_{p+m} - 2|\omega_e|)} \right) i^{|m|+|m-l|},$$

$$D^{p+l} = \left[ \sum_{j=0}^{1} \frac{ik_2 \varepsilon_{12}^l + i\alpha_j \varepsilon_{11}^l - |k_2| \delta_{0,l} \varepsilon_{\text{d}}}{(\partial\Delta/\partial\alpha_j) tg(\alpha_j a_{\text{pl}})} \right]^2$$
$$- \left[ \sum_{j=0}^{1} \frac{ik_2 \varepsilon_{12}^l + i\alpha_j \varepsilon_{11}^l + |k_2| \delta_{0,l} \varepsilon_{\text{d}}}{(\partial\Delta/\partial\alpha_j) \sin(\alpha_j a_{\text{pl}})} \right]^2 - \delta_{0,l},$$

$$F_{m,l} = G_{m,l}\left[\sum_{j=0}^{1}\frac{ctg(\alpha_j a_{\text{pl}})}{\left(\frac{\partial\Delta}{\partial\alpha_j}\right)}\cdot\sum_{j=0}^{1}\frac{ik_2\varepsilon_{11}^l - i\alpha_j\varepsilon_{11}^l - |k_2|\delta_{0,l}\varepsilon_{\text{d}}}{\left(\frac{\partial\Delta}{\partial\alpha_j}\right)tg(\alpha_j a_{\text{pl}})}\right.$$
$$\left. - \sum_{j=0}^{1}\frac{1}{\left(\frac{\partial\Delta}{\partial\alpha_j}\right)\sin(\alpha_j a_{\text{pl}})}\sum_{j=0}^{1}\frac{ik_2\varepsilon_{11}^l - i\alpha_j\varepsilon_{11}^l + |k_2|\delta_{0,l}\varepsilon_{\text{d}}}{\left(\frac{\partial\Delta}{\partial\alpha_j}\right)\sin(\alpha_j a_{\text{pl}})}\right].$$

One can express the solutions of the set of equations (4.42) for the second electron cyclotron resonance: $w = 2|w_e| + \Delta_{Te} + n\omega_0 + \gamma$, where $|\gamma| \ll 2|\omega_e|$, in the following form [62]:

$$\gamma \approx \begin{cases} -\Delta_{Te} + 1.1 \cdot i\Delta_{Te}a_E, & |k_2|a_{\text{pl}} \gg 1, \\ -\Delta_{Te}(1 + a_E^2) + 0.86 \cdot i\Delta_{Te}a_E \cdot \left(1 - 0.36k_2^2a_{\text{pl}}^2\right), & |k_2|a_{\text{pl}} \ll 1. \end{cases} \tag{4.43}$$

It follows from the analysis of the expressions (4.43), that the parametric action of the external RF field on SCX-mode propagation results in a decrease of the wave frequency. The parametric instability growth rates $\text{Im}(\gamma)$ weakly depend on the transverse dimension of the plasma layer. The value of the rates sharply decreases with increasing SCX-mode wavelength. Note that the SCX-mode damping rate caused by plasma particle collisions increases with increasing wavelength [31, 63].

The results of the numerical analysis of (4.42) are presented in Figs. 4.9, 4.10 and 4.11. In Fig. 4.9, the curve, which describes the SCX-mode growth rate $\text{Im}(\gamma)$ as a function of wavelength, crosses the curve describing the damping rate $\delta$ caused by the collisions. Right to this crossing point is $\text{Im}(\gamma) > \delta$. The position of this point is not fixed, it can move to the range of very long wavelengths ($|k_2|\rho_e \to 0$). This is associated with the fact that the Bessel functions of the first kind in the expressions of the plasma electric conductivity tensor (4.39) are decomposed into series in respect of the small argument $g$, which is proportional to the product of the SCX-mode wavenumber $|k_2|$ and the amplitude $E_0$ of the RF pumping field. That is why the inequality $g \ll 1$ can be satisfied in the range of longer wavelengths (or smaller values of $|k_2|$) for larger values of pumping field amplitudes. On the other hand, an increase of the value $E_0$ results in some reduction of the wavenumber range within which the inequality mentioned above is true.

The influence of $\vec{B}_0$ on $\text{Im}(\gamma)$ is similar to its influence on the SCX-mode frequency. Increasing the external stationary magnetic field results in an increase of the growth rate $\text{Im}(\gamma)$. As it was already mentioned above, in contrast to the case of bulk electron cyclotron waves, the development of SCX-mode parametric instability depends on the transverse dimension of the plasma layer. This dependence is presented in Fig. 4.10.

In agreement with the results found in the analytical study of the dependence of $\text{Im}(\gamma)$ on the plasma layer thickness, the numerical calculations indicate some increase of the SCX-mode parametric instability growth rate with decreasing plasma layer thickness.

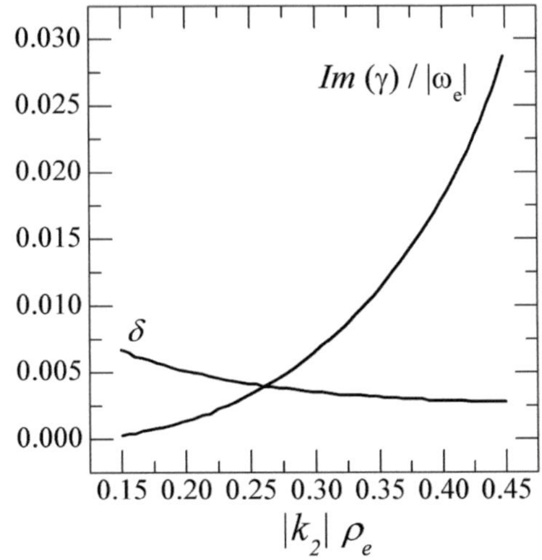


**Fig. 4.9** SCX-mode damping rate $\delta$ and parametric instability growth rate Im($\gamma$) normalized to the electron cyclotron frequency versus wavenumber multiplied by the electron Larmor radius for $\nu/|\omega_e| = 0.001$, $\beta_E = 1.2\ \rho_e$, $a_{\rm pl}/\rho_e = 0.1$, $N_e = 1000$, $\omega_0/|\omega_e| = 3.5$, $\varepsilon_{\rm d} = 1$

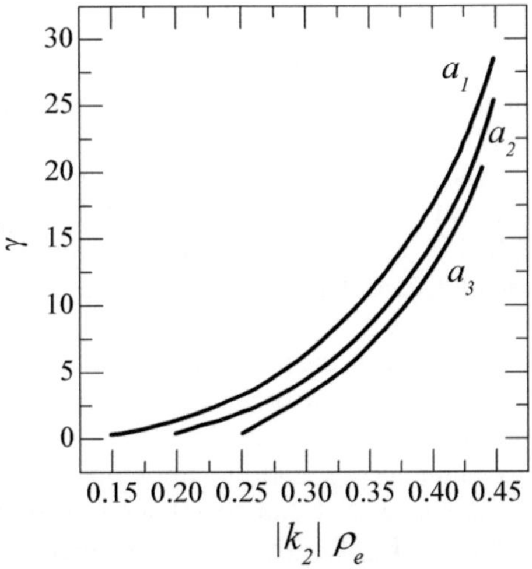


**Fig. 4.10** Dimensionless parametric instability growth rate $\gamma^{\sim} = 10^3$ Im($\gamma$)/$|\omega_e|$ versus wavenumber multiplied by the electron Larmor radius for $\beta_E = 1.2\ \rho_e$, $N_e = 1000$, $\omega_0/|\omega_e| = 3.5$, $\varepsilon_{\rm d} = 1$. The curves marked by the symbols $a_1$, $a_2$ and $a_3$ correspond to the cases $a_{\rm pl}/\rho_e = 0.1$, $a_{\rm pl}/\rho_e = 1$ and $a_{\rm pl}/\rho_e = 10$, respectively

Numerical study of the influence of the pumping field amplitude $E_0$ on Im($\gamma$) is shown in Fig. 4.11. The growth rates of SCX-mode parametric instabilities are demonstrated to increase with increasing pumping field amplitude $E_0$ and frequency $\omega_0$.

Like in the case of SCTM-mode parametric instability, the threshold value of the pumping RF field amplitude $b_{\rm th}$ can be found, for which the SCX-mode parametric instability starts to develop. The threshold can be derived via comparing the SCX-mode growth rate with its damping rate:

$$b_{\rm th} \approx \nu_e \omega^{-1} (k_2 \rho_e)^{-3}. \tag{4.44}$$

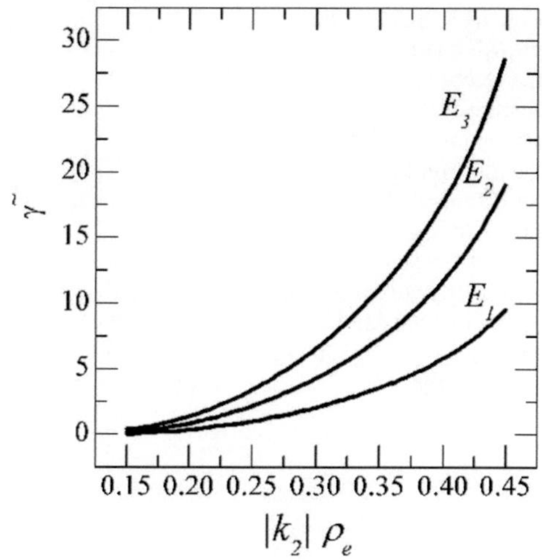


**Fig. 4.11** Dimensionless growth rate of parametric instability $\gamma^{\sim} = 10^3$ Im($\gamma$)/|$\omega_e$| versus wavenumber multiplied by the electron Larmor radius for $a_{\text{pl}}/\rho_e = 0.1, N_e = 1000,$ $\omega_0/|\omega_e| = 3.5, \varepsilon_{\text{d}} = 1$. The curves marked by the symbols $E_1$, $E_2$, and $E_3$ relate to the cases $\beta_E = 0.4\ \rho_e, \beta_E = 0.8\ \rho_e,$ and $\beta_E = 1.2\ \rho_e,$ respectively

The following conclusion comes from analysis of (4.44). To decrease the threshold value of the pumping field amplitude an experimentalist has to apply the shortest acceptable waves from the range $k_2\rho_e < 1$. Then analytical solution of the simplified (4.42) has the following approximate form:

$$\text{Im}(\gamma) \approx 2.3|\Delta_T|g_0 k_2^2 c^2/\Omega_e^2. \tag{4.45}$$

Analysis of the expression (4.45) confirms that the growth rate Im($\gamma$) of parametric instability of SCX-modes is approximately proportional to the fourth power of the wavenumber $k_2$, Im($\gamma$) $\propto k_2^4$. It increases with increasing wave penetration depth into the plasma and amplitude of external RF pumping field.

To complete the analysis of the equations which describe the initial stage of SCX-mode parametric excitation one has to solve the set (4.42) numerically. The results of this analysis are presented in Figs. 4.12 and 4.13. The dependence of the SCX-mode damping rate Im($\gamma$) on wavenumber is shown in Fig. 4.12 for different values of external magnetic field. The dimensionless parameter $N_e = \Omega_e^2/\omega_e^2$ is used there. The growth rate Im($\gamma$) increases with increasing product $k_2\rho_e$ and increasing external magnetic field $B_0$. Decreasing $B_0$ value results in increasing dimensionless parameter $N_e$. The difference between the curves calculated for $N_e > 21$ becomes very small and almost invisible, so that these curves are not shown in Fig. 4.12. This peculiarity of the dependence of Im($\gamma$) on the magnetic field $B_0$ is an interesting result, which is not obvious from the analytical expression (4.45).

The dependence of the growth rates of parametric instabilities on plasma beta, which is the dimensionless parameter $\beta = \rho_e^2\Omega_e^2 c^{-2}$ characterizing the plasma pressure, and the value of the external RF electric field amplitude $g_0$ is shown in Fig. 4.13. As it is demonstrated there, the value Im($\gamma$) increases with increasing external RF electric field amplitude and/or plasma beta, $\beta$. The curve marked by 4 in Fig. 4.13

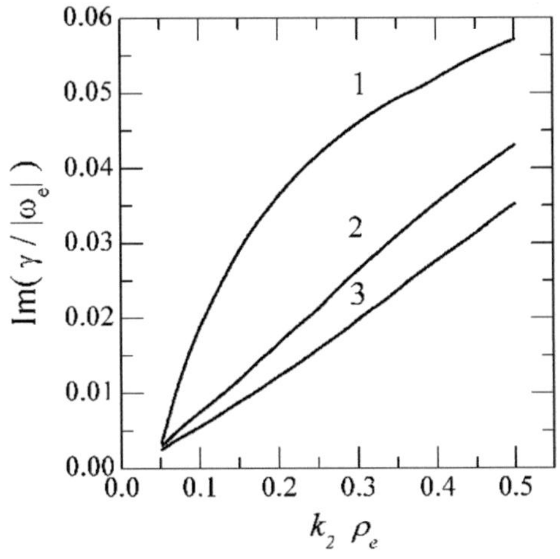


**Fig. 4.12** SCX-mode growth rate normalized to the electron cyclotron frequency, Im($\gamma$)/$|\omega_e|$, versus wavenumber multiplied by the electron Larmor radius for different values of external magnetic field for $g_0 = 0.5$, $\omega_0/|\omega_e| = 0.55$, $\beta = 0.5$. The curves marked by 1, 2, and 3 are calculated for the cases $N_e = 2$, 3, and 21, respectively

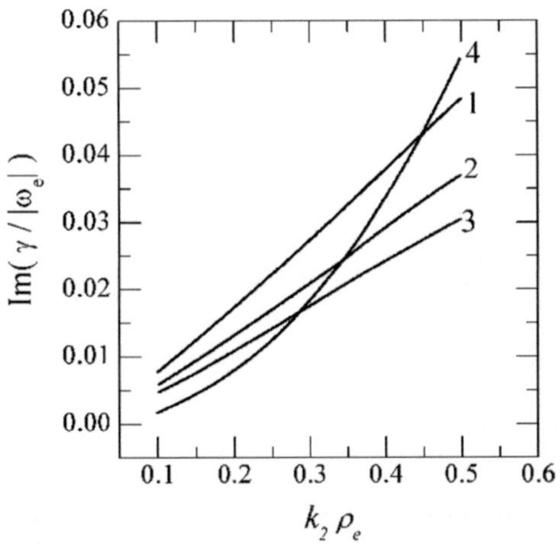


**Fig. 4.13** SCX-mode growth rate normalized to the electron cyclotron frequency, Im($\gamma$)/$|\omega_e|$, versus wavenumber multiplied by the electron Larmor radius for $\omega_0/|\omega_e| = 0.55$, $N_e = 21$. The curves marked by 1, 2, and 3 are obtained from numerical solving (4.42) for $g_0 = 0.6$, $\beta = 0.9$; $g_0 = 0.45$, $\beta = 0.9$; $g_0 = 0.45$, $\beta = 0.5$, respectively. The curve marked by 4 demonstrates good agreement of the analytical solution (4.45) with the case 2: $g_0 = 0.45$, $\beta = 0.9$

presents the results of an analytical solution, in other words, it is calculated according to (4.45). Since this analytical result is by the factor of 1.4 larger than the numerical one in the range $k_2\rho_e > 0.35$, and is smaller for $k_2\rho_e < 0.35$, one can state good agreement between the analytical and numerical solutions.

Thus, in this section, an infinite set of equations which describe the initial stage of SCX-mode parametric instability is obtained. It relates to the wave electric field of SCX-modes at the interface of plasma layer and dielectric in a dielectrically lined waveguide. The dependence of the SCX-mode growth rate on the parameters of the plasma waveguide and the pumping RF wave is analyzed. The parametric instability growth rate is shown to weakly depend on plasma particle density and pumping wave frequency. The growth rate increases with decreasing plasma layer thickness and increasing wavenumber and pumping wave amplitude.

## 4.5 Parametric Excitation of Surface Electron Cyclotron X-Modes in Non-monochromatic External RF Fields

Application of powerful generators of electromagnetic waves for experimental studies of plasmas actualizes the problem of parametric instability of the waves which are eigenwaves of the studied waveguide structure wherein the plasma is produced and confined. Generators, which work on harmonics of the electron cyclotron frequency, are often used for these purposes and the external RF field is often non-monochromatic. That is why the theoretical study of parametric excitation of slow extraordinarily polarized electromagnetic surface waves at the harmonics of the electron cyclotron frequency influenced by non-monochromatic pumping waves is an actual problem.

In the previous sections, these surface waves are shown to be eigenwaves of waveguide structures composed of a dielectrically coated metal waveguide with magnetoactive plasma filling. They propagate under the condition of weak plasma spatial dispersion. Their penetration depth into the plasma is of the order of the wavelength. Parametric instability of these waves under influence of a monochromatic pumping wave was studied in the previous Sect. 4.4. The case of non-monochromatic pumping RF field is known from the theory of bulk wave parametric instability to be different from that of monochromatic pumping. That is why the influence of the non-monochromatic external pumping wave on SCX-mode excitation is studied in this section.

The following model of the waveguide is under consideration (see Fig. 4.3). The semi-bounded uniform plasma occupies the half-space $x > 0$. It is bordered by a dielectric with thickness $a_{\mathrm{d}}$ and dielectric constant $\varepsilon_{\mathrm{d}}$ at the plane $x = 0$. The interface between dielectric and waveguide metal wall is at the plane $x = -a_{\mathrm{d}}$. The external stationary magnetic field $\vec{B}_0$ is oriented along the $\vec{z}$ axis. The model of semibounded plasma is applicable if the plasma transverse dimension is much larger than the wave penetration depth into the plasma. As it is shown in Sect. 4.3, the SCX-mode dispersion properties weakly depend on the transverse plasma dimension. Here the model of a uniform plasma particle density is chosen since plasma particle density inhomogeneities weakly influence the frequency of surface electron cyclotron waves of this polarization (see Sect. 4.3).

The motion of plasma particles is described by the kinetic Vlasov–Boltzmann equation. An unperturbed Maxwellian distribution function is chosen. The space is assumed to be uniform along the $\vec{z}$ axis, which means $\partial/\partial z = 0$. The case of weak plasma spatial dispersion is considered so that $\left|\vec{k}\rho_\alpha\right| \ll 1$. The model of diffusive reflection of plasma particles from the plasma interface is applied. In this case, the components of the plasma permittivity tensor contain also the anti-Hermitian terms, which determine surface wave kinetic damping. These terms are omitted in this section since the latter deals with SCX-mode excitation.

Solving the kinetic equation by the method of trajectories one derives the following expressions for the Fourier coefficients $j_1$ and $j_2$ of the electric currents $j_x$ and $j_y$, respectively:

$$
\begin{aligned}
j_1(n_1,n_2) = &\sum_{s=-\infty}^{+\infty}\sum_{m_1=-\infty}^{+\infty}\sum_{m_2=-\infty}^{+\infty}\sum_{u_1=-\infty}^{+\infty}\sum_{u_2=-\infty}^{+\infty}[\sigma_1(s,m_1,m_2,u_1,u_2,n_1,n_2) \\
&\times E_1(m_1,m_2) + i\sigma_2(s,m_1,m_2,u_1,u_2,n_1,n_2)E_2(m_1,m_2)], \\
j_2(n_1,n_2) = &\sum_{s=-\infty}^{+\infty}\sum_{m_1=-\infty}^{+\infty}\sum_{m_2=-\infty}^{+\infty}\sum_{u_1=-\infty}^{+\infty}\sum_{u_2=-\infty}^{+\infty}[\sigma_3(s,m_1,m_2,u_1,u_2,n_1,n_2) \\
&\times E_2(m_1,m_2) - i\sigma_2(s,m_1,m_2,u_1,u_2,n_1,n_2)E_1(m_1,m_2)],
\end{aligned}
\tag{4.46}
$$

$$
\begin{aligned}
&\sigma_3(s,m_1,m_2,u_1,u_2,n_1,n_2) \\
&\approx \sigma_1(s,m_1,m_2,u_1,u_2,n_1,n_2) \\
&= \frac{\Omega_e^2 s^2 I_s(y_e) J_{u_1-m_1}(a_1) J_{u_1-n_1}(a_1) J_{u_2-m_{21}}(a_2) J_{u_2-n_2}(a_2)}{4\pi i y_e(s\omega_e - \omega - u_1\omega_{01} - u_2\omega_{02})\exp(y_e)}; \\
&\sigma_2(s,m_1,m_2,u_1,u_2,n_1,n_2) \\
&= \frac{\Omega_e^2 s\left[I_s'(y_e) - I_s(y_e)\right]}{4\pi i y_e(s\omega_e - \omega - u_1\omega_{01} - u_2\omega_{02})\exp(y_e)} \\
&\quad\times I_s(y_\alpha) J_{u_1-m_1}(a_1) J_{u_1-n_1}(a_1) J_{u_2-m_{21}}(a_2) J_{u_2-n_2}(a_2);
\end{aligned}
\tag{4.47}
$$

here $y_e = k_1^2\rho_e^2/2$, $k_1$ is the component of the wave vector along the $\vec{x}$ axis, $I_s'(z)$ is the derivative of the modified Bessel function with respect to its argument, and the parameters $a_n = |e|k_1E_{0n}/\left[m_\alpha\left(\omega_{0n}^2 - \omega_e^2\right)\right]$ characterize the amplitudes of the external electric pumping field components. The latter is chosen in the following two-frequency form:

$$
\vec{E}_0(t) = \vec{E}_{01}\sin(\omega_{01}t) + \vec{E}_{02}\sin(\omega_{02}t),
\tag{4.48}
$$

where $\vec{E}_{0i}$ are the amplitudes of the external electric field components with the frequencies $\omega_{0i}$. To simplify the calculations it is assumed that $\vec{E}_{0i} \parallel \vec{x}$.

The fields of SCX-modes at the second harmonic of the electron cyclotron frequency are assumed to depend on time and coordinates in the following manner:

$$
\left\{\begin{matrix} E_{x,y}(t,y) \\ H_z(t,y) \end{matrix}\right\} \sim \sum_{n_1=-\infty}^{+\infty}\sum_{n_2=-\infty}^{+\infty}\left\{\begin{matrix} E_{1,2}(n_1,n_2) \\ H_3(n_1,n_2) \end{matrix}\right\}\exp\left(ik_2y - i\omega_{n_1+n_2}t\right),
\tag{4.49}
$$

where $\omega_{n_1+n_2} = \omega + n_1\omega_{01} + n_2\omega_{02}$.

Here the case of sufficiently slow waves is under consideration for which $|\sigma_{1,2}|\upsilon_{\text{ph}}^2 \ll \omega_{n_1+n_2}c^2$, where $\upsilon_{\text{ph}}$ is the phase velocity of these waves. From the Maxwell equations, the following equation can be derived for the Fourier coefficient of the tangential component of the SCX-mode:

$$cH_z(x=0,n_1,n_2)\left(2i\pi\omega_{n_1+n_2}\right)^{-1} + 4i\pi\, j_2(n_1,n_2)(k_2+k_1)k_2^{-1}\omega_{n_1+n_2}^{-1}$$
$$= -E_2(n_1,n_2)\left(1+k_1^2\omega_{n_1+n_2}^2c^{-2}k_2^{-4}\right), \tag{4.50}$$

where $H_z(x=0)$ is the value of the magnetic RF wave field at the plasma interface.

The boundary conditions for the fields of the SCX-mode in this case are similar to those applied in the previous sections. However, one condition differs from those used in Sect. 4.4, namely: a nonlinear electric current is present at the plasma-dielectric interface caused by the action of the external RF field. This boundary condition makes it possible to determine the discontinuity of the SCX-mode magnetic field at this interface. The boundary condition can be written in the following form:

$$H_z(x=0,n_1,n_2)$$
$$= \sum_{m_1,m_2=-\infty}^{+\infty}\sum_{l_1,l_2=-\infty}^{+\infty} \frac{-i(ib_1\rho_e k_2/2)^{|m_1|+|m_1-l_1|}(ib_2\rho_e k_2/2)^{|m_2|+|m_2-l_2|}}{ck_2 m_1!(m_1-l_1)!m_2!(m_2-l_2)!}$$
$$\times\left[\frac{\omega_e\Omega_e^2}{\left(\omega_{n_1+m_1+n_2+m_2}\right)^2-\omega_e^2}+\frac{\Omega_e^2 y_e}{\omega_{n_1+m_1+n_2+m_2}h_e}\right]$$
$$\times E_y(x=0,n_1+l_1,n_2+l_2), \tag{4.51}$$

where $b_j = a_j(k_e\rho_i)^{-1}$. To derive the explicit form of the boundary condition (4.51) the assumption of small pumping wave amplitude is used: $|\rho_e b_j k_2/2| \ll 1$.

Inverse Fourier transformation of (4.50) makes it possible to derive the following infinite set of equations for the $n_1, n_2$-th harmonics of the tangential electric field $E_y(x = 0, n_1, n_2)$ at the plasma–dielectric interface:

$$D(n_1,n_2)E_y(x=0,n_1,n_2) + F(n_1,n_2,l_1,l_2)E_y(x=0,n_1,n_2,l_1,l_2) = 0, \tag{4.52}$$

where

$$D(n_1,n_2) = \frac{i|k_2|q_0}{q_0^2+k_2^2}\left[\varepsilon_1\left(\omega_{n_1+n_2}\right)+\varepsilon_2\left(\omega_{n_1+n_2}\right)-\varepsilon_{\text{d}}\coth|a_{\text{d}}k_2|\right]+\varepsilon_1\left(\omega_{n_1+n_2}\right)$$
$$-\varepsilon_2\left(\omega_{n_1+n_2}\right)+\varepsilon_{\text{d}}\coth|a_{\text{d}}k_2|,\quad N_e=\Omega_e^2\omega_e^{-2}$$

$$F(n_1,n_2,l_1,l_2) = \sum_{l_1,l_2=-\infty}^{+\infty} \sum_{m_1,m_2=-\infty}^{+\infty} \left[ \frac{\omega_e^3}{\left(\omega_{n_1+m_1+n_2+m_2}\right)^2 - \omega_e^2} - \frac{\omega_e^2 \left(k_2^2 \rho_e^2/2\right)^2}{\omega_{n_1+m_1+n_2+m_2} h_e} \right]$$
$$\times \frac{4N_e (ib_1 \rho_e k_2/2)^{|m_1|+|m_1-l_1|} (ib_2 \rho_i k_2/2)^{|m_2|+|m_2-l_2|}}{m_1!(m_1-l_1)!m_2!(m_2-l_2)!\omega_{n_1+n_2}};$$
$$q_0 = \frac{4i}{\rho_e}\sqrt{\frac{h_e}{3}}.$$

This set of equations transforms to the set (4.42) in the limiting case of monochromatic external RF field if in (4.52) amplitude and frequency of one of the pumping waves is taken to be equal to zero, e.g.: $b_1 \to 0$ and $\omega_{01} \to 0$.

In the general case, when neither $\omega_{01}$ nor $\omega_{02}$ are equal to zero, there is a possibility to study the set (4.52) analytically if $\omega_{01}$ and $\omega_{02}$ relate to each other as natural numbers $p_1$ and $p_2$ so that $\omega_{01} = \omega_{02} p_1/p_2$ (see, e.g., [64, 65]). In the case of such relation between the frequencies, the problem can be reduced to the case of parametric instability caused by action of a monochromatic pumping wave with effective frequency $\omega_0 = (\omega_{01} - \omega_{02})/(p_1 - p_2)^{-1}$. Then the sum $n_1\omega_{01} + n_2\omega_{02}$ can be replaced by $N_0\omega_0$, where this specific $N_0$ can be determined for each two numbers $n_1'$ and $n_2'$, namely $N_0 = n_1' p_1 + n_2' p_2$. This replacement results in an algebraic set of equations for $E_y(x=0, N, l_1, l_2)$, which has nontrivial solutions if the determinant composed from the coefficients of the considered harmonics of the SCX-mode tangential electric field is equal to zero.

Below, the case of the second electron cyclotron frequency is investigated: $\omega = 2|\omega_e| + 0.5\omega_e k_2^2 \rho_e^2 + N_0\omega_0 + \gamma$. Keeping the main and two closest harmonics of the SCX-mode, one can derive the following equation, which makes it possible to determine the growth rates of parametric instability at the initial stage of its development [66]:

$$\begin{vmatrix} D(-1); & F(-1;+1); & F(-1;+2) \\ F(0;-1); & D(0); & F(0;+1) \\ F(+;-2); & F(+1;-1); & D(+1) \end{vmatrix} = 0. \tag{4.53}$$

In (4.53), for simplicity it is assumed that $N_0 = 0$. Analysis of (4.53) gives principally different results for the following two correlations between the frequencies of the pumping waves and the electron cyclotron frequency.

First, if neither $\omega_{01}$ nor $\omega_{02}$ is close to the electron cyclotron frequency, (4.53) can be reduced to an algebraic quadratic equation regarding $\gamma$, which has no complex roots. Then the SCX-mode parametric instability does not develop under these conditions.

Second, if one of the frequencies of the pumping waves, e.g., $\omega_{01}$ is close to the electron cyclotron frequency, (4.53) can be reduced to an algebraic cubic equation regarding $\gamma$. One solution makes it possible to calculate the maximum magnitude of the growth rate of the parametric instability under the condition of relatively small values of the pumping wave amplitudes:

$$\mathrm{Im}(\gamma) \approx 0.5|\omega_e|k_2^2\rho_e^2\sqrt[3]{b_1(b_1+0.7b_2)}. \tag{4.54}$$

This expression differs from that obtained in the case of monochromatic pumping wave. The maximum value of the growth rate (4.54) is larger than that obtained in the previous section given by (4.45).

Thus, unlike in the case of monochromatic pumping wave, in the present case there are two different regimes of action of non-monochromatic pumping waves on SCX-modes. If one of the frequencies of the pumping waves is close to the electron cyclotron frequency, the regime of surface waves excitation is realized at the second harmonic of the electron cyclotron frequency. If no frequency is close to $2\omega_e$, no excitation of SCX-modes takes place [67].

The scenario of the parametric instability strongly depends on the correlation between the frequencies $\omega_{01}$ and $\omega_{02}$ of the pumping waves, which results in a non-monotonous dependence of the growth rate $\mathrm{Im}(\gamma)$ on the pumping wave amplitudes. That is why more detailed information on the parametric excitation should be obtained from numerical analysis of (4.52). The results of these studies are presented in Figs. 4.14 and 4.15.

The dependence of the growth rate $\mathrm{Im}(\gamma)$ on the SCX-mode wavenumber is shown in Fig. 4.14 for different values of plasma waveguide parameters. The influence of the dielectric constant $\varepsilon_d$ and thickness $a_d$ of the protective dielectric coating on $\mathrm{Im}(\gamma)$ is negligible as compared with that of the external static magnetic field and the pumping wave amplitudes. Increase of $\varepsilon_d$ and decrease of $a_d$ result in some increase of the SCX-mode growth rate. Increasing external magnetic field also increases $\mathrm{Im}(\gamma)$, but much more significantly.

As shown in Fig. 4.15, the main influence on $\mathrm{Im}(\gamma)$ (for fixed values of the other plasma waveguide parameters) is caused by the amplitude of the second (non-resonant) pumping wave whose frequency $\omega_{02}$ is far from the electron cyclotron

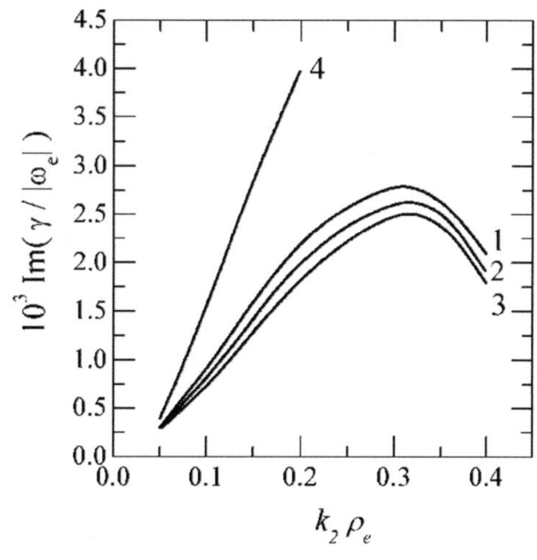


**Fig. 4.14** SCX-mode growth rate versus wavenumber for $b_1 = 0.8$, $b_2 = 0.1$, $\omega_{01} = 0.99\ |\omega_e|$, $\omega_{02} = 0.27\ |\omega_e|$, $\omega_0 = 0.09\ |\omega_e|$. The curves marked by the numbers 1, 2, and 3 correspond to the cases: $\varepsilon_d = 4$, $a_d = 100\ \rho_e$, $N_e = 100$; $\varepsilon_d = 4$, $a_d = 100\ \rho_e$, $N_e = 500$; and $\varepsilon_d = 1$, $a_d = 100\ \rho_e$, $N_e = 500$, respectively. The line marked by 4 is calculated from the analytical expression (4.54) for $\mathrm{Im}(\gamma)$

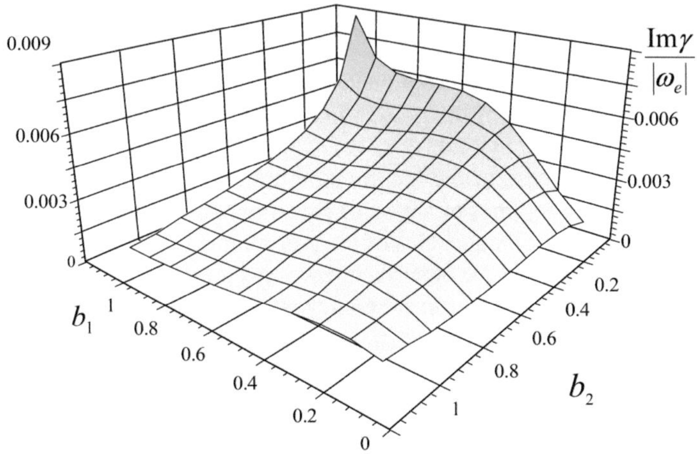


**Fig. 4.15** SCX-mode growth rate, normalized to the electron cyclotron frequency, versus amplitudes of external pumping electric fields for $\varepsilon_\mathrm{d} = 4$, $a_\mathrm{d} = 100\ \rho_e$, $N_e = 500$, $|k_2\ \rho_e| = 0.25$. The frequencies of the pumping fields are the same as in Fig. 4.14

resonance. One can see that increasing wavenumber first causes an increase of $\mathrm{Im}(\gamma)$, then the growth rate achieves its maximum value, and starts to decrease. If the amplitude of the pumping wave with the frequency $\omega_{01} \approx \omega_e$ is sufficiently large ($b_1 \geq 0.5$), increasing amplitude $b_2$ of the second pumping field (whose frequency $\omega_{02} \neq \omega_e$) results in almost monotonous decrease of the growth rate $\mathrm{Im}(\gamma)$. However, if $b_1 < 0.5$, an increase of $b_2$, first results in increasing growth rate and then to its decrease.

Thus, in this section it is found, that a non-monochromatic spectrum of the pumping waves changes the scenario of parametric instability as compared to the case of monochromatic excitation of SCX-modes. The present results make it possible to conclude that the main influence on the growth rate of SCX-mode parametric instability is caused by the correlation between the frequencies of the pumping fields. If one of these frequencies is close to a harmonic of the electron cyclotron resonance frequency $\omega_e$, a large value of SCX-mode growth rate becomes possible. The second important factor is the value of the amplitude of the second, non-resonant frequency of the pumping RF field. Changing the amplitude one can control the SCX-mode growth rate more efficiently than changing the parameters of the plasma itself. This means that application of a second RF generator with a frequency far from the harmonic of the electron cyclotron resonance frequency can be an efficient tool for controlling the parametric instability development of SCX-modes.

## 4.6 Gas Discharges Sustained by Surface Electron Cyclotron X-Modes

This section is devoted to a theoretical analysis of gas discharges sustained by SCX-mode waves. These waves are eigenwaves of the discharge chamber, which can be modelled by a planar waveguide with metal walls, protective coating, and plasma filling. The external static magnetic field is oriented along the plasma interface. The dispersion properties of surface cyclotron X-modes at the harmonics of the electron cyclotron frequency which propagate across the external magnetic field in such waveguide structures are studied in Sect. 4.3. The phase velocity of SCX-modes weakly depends on the plasma particle density and is mainly determined by the value of the external magnetic field which seems to be convenient in view of the possibility to control the discharge parameters.

The geometry of the discharge chamber is the same as in the previous Sect. 4.5. The planar discharge structure is realized by a planar metal waveguide whose wall is located in the region $x \leq -a_{\rm d}$. The plane $x = -a_{\rm d}$ can be used to carry an RF electric current in order to excite electromagnetic waves in the plasma. To protect the plasma from contamination by impurity ions from the wall material, the internal interface of the waveguide is covered by a dielectric coating with the dielectric constant $\varepsilon_d$. The dielectric occupies the layer $0 > x > -a_{\rm d}$. The plasma is placed in the half-space $x \geq 0$. The vector of the external stationary magnetic field $\vec{B}_0 \parallel \vec{z}$ is parallel to the plasma interface. The stationary gas discharge is assumed to be sustained by a slow SCX-mode. The discharge space is supposed to be uniform along the $\vec{z}$ axis. The thermal motion of the plasma particles is assumed to be weak so that the following inequality is valid: $\left|\vec{k}\right| \rho_e \ll 1$. The plasma particle density is supposed to be sufficiently large: $\Omega_e^2 \gg \omega_e^2$. This case is the most interesting from the point of view of practical applications [68–73]. As found earlier, SCX-mode damping takes place due to two mechanisms. Plasma particle collisions are the reason for the first mechanism. They can be described in a phenomenological way via introducing the effective collision frequency $\nu_e$ (in this case the inequality $\omega \gg \nu_e$ is assumed). The second mechanism is resonant wave–particle interaction.

In Sect. 4.3, SCX-modes are shown to propagate along the plasma–dielectric interface with both signs of the wavenumber: $k_2 < 0$ and $k_2 > 0$ [37]. If the thickness of the dielectric coating vanishes ($a \to 0$), only one mode with negative wavenumber ($k_2 < 0$) remains [74]. That is why the following consideration is restricted to the case of SCX-modes with negative wavenumbers. SCX-modes at the second and fourth harmonics of the electron cyclotron frequencies are studied here.

The initial equations in this problem are the kinetic equation for the plasma particles and the Maxwell equations for the wave fields. In the framework of nonlocal kinetic approach [75–82] the energy dependence of the electron distribution function can be divided into three main ranges of the energy $W$. The first range includes small values of energy which are smaller than the lowest level $\Theta^*$ of plasma particle excitation ($W < \Theta^*$). Within this range the model of elastic collisions can be applied to describe the particle interaction. The second range corresponds to the energies, which are larger than $\Theta^*$. The particles interact inelastically within this range. Finally, the third range of energies describes suprathermal electrons with $W \gg \Theta^*$.

An absolute majority of plasma electrons in microwave gas discharges belongs just to the first and second energy ranges. Only an insignificant part of electrons has an energy from the third range—they are referred as high-energy tail of the electron distribution function. The electron distribution function $f$ is almost isotropic for the electrons from the first and second energy ranges [76–82] however, it becomes anisotropic only due to existence of suprathermal electrons. That is why one can apply the following approximate description for the distribution function of plasma electrons in the gas discharge: $f = f_0 + f_1$, wherein the following inequality for the isotropic $f_0$ and anisotropic $f_1$ components is assumed: $f_0 \gg f_1$. Application of the nonlocal kinetic approach is justified in the case when electron diffusion in the discharge space takes place faster than the electron energy change due to their collisions with the plasma particles. That is why the anisotropy of the electron distribution function is small, and the total electron energy is approximately invariant of the motion. Therefore, one can derive the following equation for the anisotropic component of the electron distribution function:

$$\frac{\partial}{\partial t} f_1 + (\vec{\upsilon}, \nabla) f_1 - \omega_\alpha \frac{\partial}{\partial \varphi} f_1 + \frac{e_\alpha \vec{E}}{m_\alpha} \frac{\partial f_0}{\partial \vec{\upsilon}} = -\nu_\alpha f_1, \tag{4.55}$$

where the subscript $\alpha$ indicates the plasma particle species, $\varphi$ is the angle in the velocity-phase space, and the integral of collisions is written in the framework of the model of effective collision frequency: $(-\nu_\alpha f_1)$ [26–28, 35, 36]. Since some part of electrons interact with the plasma interface, one has to take this interaction into account. For this purpose, one can apply the model of plasma particle diffusive reflection from the interface. This model is widely used for theoretical description of the phenomena nearby the plasma interface [25, 29, 34]. Under these conditions one term more appears in the expression for the Fourier coefficients of the electric current:

$$j_l = \sigma_{lm}(k_1) E_m(k_1) + \sigma_{lm}^{(ah)}(k_1) E_m(k_1), \tag{4.56}$$

where $\sigma_{ik}(k_1)$ is the tensor of electric conductivity of an infinite magnetoactive plasma in the approach of weak plasma spatial dispersion and $\sigma_{ik}^{(ah)}(k_1)$ is the anti-Hermitian part of the plasma conductivity tensor—its appearance is caused just by plasma particle interaction with the plasma interface. Its magnitude determines the kinetic SCX-mode damping.

In the approach of weak plasma spatial dispersion applied in this section one can use the asymptotic expressions [see e.g. (3.18)] for $\sigma_{ik}^{(ah)}$ from [25, 34]. Since ion terms are much smaller than corresponding electron terms for waves at harmonics of the electron cyclotron frequency, the contribution of ion terms is neglected to simplify the form of the components of the tensor $\sigma_{ik}^{(ah)}$.

SCX-mode energy losses caused by neutral gas ionization are described by the energy balance equation. In the present case, this equation has the following form:

$$dS_y/dy = -Q = -Q_{col} - Q_{kin}. \tag{4.57}$$

The value $Q$ is determined by the wave energy losses caused by Ohmic dissipation and resonant wave–particle interaction:

$$Q = \frac{\omega}{8\pi}\int_0^\infty \mathrm{Im}(\varepsilon_1)|E(x)|^2 dx + \frac{1}{2}\int_0^\infty \sigma_{ik}^{(ah)} E_k(x)E_i(x)dx = Q_{oh} + Q_{kin}, \tag{4.58}$$

where $\mathrm{Im}(\varepsilon_1) = \nu_e\Omega_e^2(\omega^2+\omega_e^2)/\left[\omega(\omega^2-\omega_e^2)^2\right]$, and $E(x)$ is the SCX-mode electric field in the plasma region.

The SCX-mode frequency at the second harmonic of the electron cyclotron frequency has the following form:

$$\omega \approx |\omega_e|\left[2+\zeta(\eta)k_2^2\rho_e^2\right], \tag{4.59}$$

where $\eta = \mathrm{sgn}(k_2)$, $\zeta(1) \approx 1$, $\zeta(-1) \approx 0.034$, and $|k_2|\rho_e \ll 1$. Here the case of negative wavenumbers is considered ($k_2 < 0$), since SCX-modes do not propagate with positive wavenumbers in the present waveguide structures if the protective dielectric coating is sufficiently thin [33].

The following expressions for the SCX-mode fields in the region occupied by the plasma can be found from Maxwell equations by the method of Fourier analysis:

$$\begin{aligned}
E_x(x) &\approx -\frac{iE_y(0)}{2\varepsilon_1}\Phi\left(e^{-x|k_2|}+\frac{q^2}{1-q^2}e^{-x|k_2|q}\right),\\
E_y(x) &\approx \frac{E_y(0)}{2\varepsilon_1}\Phi\left(e^{-x|k_2|}+\frac{q^2}{1-q^2}e^{-x|k_2|q}\right),\\
H_z(x) &\approx \frac{kiE_y(0)}{2\varepsilon_1|k_2|}\Phi\left[(\varepsilon_2-\varepsilon_1)e^{-x|k_2|}+\frac{q(\varepsilon_1+\varepsilon_2)}{1-q^2}e^{-x|k_2|q}\right],
\end{aligned} \tag{4.60}$$

where $\Phi = \varepsilon_1+\varepsilon_2-\varepsilon_d\coth(|k_2|a_d)$, $q = \sqrt{-16\varepsilon_1\cdot h_e/(k_2^2\rho_e^2N_e)}$ is a root of the dispersion relation (4.21), $E_y(0)$ is the value of tangential component of the SCX-mode electric field at the plane $x = 0$, $N_e = \Omega_e^2/\omega_e^2$, $h_e = 1-2\omega_e/\omega$, and $\varepsilon_1$ and $\varepsilon_2$ are the components of the plasma permittivity tensor in hydrodynamic approach.

From (4.58), the following expression can be obtained for the SCX-mode energy absorbed by the plasma through the channel of Ohmic damping:

$$Q_{\text{oh}} \approx \frac{\Omega_e^2(\omega^2+\omega_e^2)\nu_e E_y^2(0)}{16\pi\cdot|k_2|(\omega^2-\omega_e^2)}\left[1+\frac{(q-1)^2}{8q}\left(1-\frac{\varepsilon_2}{\varepsilon_1}+\frac{\varepsilon_d}{\varepsilon_1\tanh(|k_2|a_d)}\right)^2\right]. \tag{4.61}$$

As it follows from the analysis of expression (4.61), the value of Ohmic energy losses is a little bit larger in the case of thin dielectric layer ($|\varepsilon_1| \ll \coth(a_{\text{d}}|k_2|\varepsilon_{\text{d}})$), than in the case of a thick layer. Increasing SCX-mode wavelength results in increasing $Q_{\text{oh}}$. The magnetic component $|H_z|$ of SCX-modes is much smaller than the electric components which are of the same order $|E_x|$~$|E_y|$ in contrast to the case of SCTM-modes (see Sect. 3.7). This is associated with the fact that SCX-mode waves are sufficiently slow (the phase velocities satisfy the inequality: $\upsilon_{\text{ph}}^2\ |\varepsilon_1| \ll c^2$) and extraordinarily polarized.

The average value of the Poynting vector for SCX-modes in the considered discharge chamber can be found from calculating the following integral:

$$S_y = -\frac{c}{8\pi}\left[\int_0^{\infty} E_x^P H_z^P dx + \int_{-a_{\text{d}}}^{0} E_x^D H_z^D \text{d}x\right], \tag{4.62}$$

where the superscript $D$ indicates that the SCX-mode fields are calculated in the region occupied by the dielectric coating of the discharge chamber. The energy flow can be written in the form of the sum $S_y = S_y^P + S_y^D$. However, since the following inequality is valid: $|S_y^D| \ll |S_y^P|$, the value $S_y^D$ can be neglected. The explicit expressions (4.60) for the electron SCX-mode fields make it possible to calculate the integral in (4.62) for the energy flow of this wave at the second harmonic of the electron cyclotron frequency:

$$S_y^P \approx \frac{\omega E_y^2(0)}{32\pi k_2^2\varepsilon_1}\left[\varepsilon_2^2-\varepsilon_1^2+\frac{2\varepsilon_{\text{d}}}{\tanh(|k_2|a_{\text{d}})}\left(\frac{0.5\varepsilon_{\text{d}}}{\tanh(|k_2|a_{\text{d}})}+\varepsilon_2+\varepsilon_1\frac{q+1}{q-1}\right)\right]\frac{q-1}{q+1}. \tag{4.63}$$

Simple analytic expressions for the values $Q$ and $S_y^P$ can be derived in the limiting cases of thick ($|\varepsilon_1,\ \varepsilon_2| \gg \varepsilon_{\text{d}}\cdot\coth(|k_2|a_{\text{d}})$) and thin ($|\varepsilon_1,\ \varepsilon_2| \ll \varepsilon_{\text{d}}\cdot\coth(|k_2|a_{\text{d}})$) dielectric coating of the discharge chamber wall. In the case of thick dielectric layer, the expression for the electron SCX-mode energy flow can be written in the following form:

$$S_y^P \approx 0.16N_e\omega E_y^2(0)/\left(\pi\ k_2^2\right). \tag{4.64}$$

In the opposite limiting case of thin dielectric layer, the expression for $S_y^P$ is given by:

$$S_y^P \approx 0.014\varepsilon_{\text{d}}^2\coth^2(a_{\text{d}}k_2)\omega E_y^2(0)/\left(\pi\cdot k_2^2|\varepsilon_1|\right). \tag{4.65}$$

Comparison of the expressions (4.64) and (4.65) for the electron SCX-mode energy flow leads to the conclusion that a reduction of the thickness of the protective dielectric coating results in an increase of the energy flow $S_y^P$ due to the enhanced factor coth( $|k_2|\, a_\mathrm{d}$).

One can derive the following simple first order differential equation for the plasma particle density from the energy balance equation (4.57):

$$d\ln(n)/dy = -1/L, \tag{4.66}$$

where the parameter $L$ is the discharge length. Then the plasma particle density profile along the direction of electron SCX-mode propagation is described by the following expression in the case of Ohmic wave energy transfer to the plasma:

$$n(y) \approx n(0)(1 - y/L), \tag{4.67}$$

where

$$L \approx |\omega_e|\left(2 + 0.034k_2^2\rho_e^2\right)/(\varsigma\nu_e|k_2|). \tag{4.68}$$

The factor $\varsigma$ is approximately equal to 2.5 in the case of thin dielectric layer, and to 10 in the opposite limiting case of thick dielectric layer.

The collision frequency between the plasma particles drastically decreases under the condition of low pressure of the working gas in RF discharges. Then the Ohmic mechanism of energy transfer becomes ineffective. However, there are two more mechanisms of wave energy transfer to the plasma for surface type waves. One of them is the kinetic mechanism, which is caused by the resonant plasma particle–wave interaction. The value of the energy, which is transferred through this channel, does not depend on the collision frequency. Then, under the condition of low pressure of the working gas, the energy absorbed due to Ohmic dissipation is much smaller than that absorbed through the channel of kinetic dissipation of surface wave energy.

In Sect. 3.7, the kinetic wave–particle interaction is shown to become the dominant mechanism of the transfer of SCTM-mode energy to the plasma under the condition of low pressure of the working gas. This initiates calculating the wave energy kinetic losses in the present case as well. One-sided Laplace transformation makes it possible to derive the following expression for the electron SCX-mode energy, which is resonantly absorbed by the plasma:

$$\begin{aligned} Q_\mathrm{kin} = \frac{c}{8}E_y^2(0)\left[\sigma_{11}^{(\mathrm{ah})}\left(\frac{\varepsilon_1+\varepsilon_2-\varepsilon_\mathrm{d}\coth(a_\mathrm{d}|k_2|)}{2\varepsilon_1\left(1-q^2\right)}\right)^2 + \sigma_{22}^{(\mathrm{ah})}\right] \\ + i\sigma_{12}^{(\mathrm{ah})}\frac{\varepsilon_1+\varepsilon_2-\varepsilon_\mathrm{d}\coth(a_\mathrm{d}|k_2|)}{\varepsilon_1\left(1-q^2\right)8}cE_y^2(0). \end{aligned} \tag{4.69}$$

The expression (4.69) can be simplified in the limiting case of thick protective dielectric layer ($|a_d k_2 \gg 1|$) to get the following asymptotic expression for $Q_{kin}$:

$$Q_{kin} \approx 0.41\pi^{-3/2} \upsilon_{Te} N_e E_y^2(0). \tag{4.70}$$

In the case $|a_d k_2 \ll 1|$, which means that the protective dielectric layer is thin, one derives the following asymptotic expression for $Q_{kin}$:

$$Q_{kin} \approx 0.015\pi^{-3/2} \upsilon_{Te} N_e \varepsilon_d^2 \coth(a_d|k_2|) E_y^2(0). \tag{4.71}$$

The ratio $Q_{kin}/Q_{oh}$ can be estimated as $\omega|k_2|\rho_e \nu_e^{-1}$ and $0.1 \cdot \omega|k_2|\rho_e N_e^2 \nu_e^{-1}$, respectively, for the cases of thick and thin dielectric coating. Thus, one can uniquely conclude that the kinetic mechanism of electron SCX-mode energy transfer to plasmas dominates over Ohmic heating in the case when the following inequality is realized: $\omega|k_2|\rho_e > \nu_e$.

The discharge length $L_{kin}$ can be derived from the energy balance equation (4.57), expressions (4.64) and (4.65) for the electron SCX-mode energy flow, and (4.70) and (4.71) for the wave energy, which can be absorbed by the plasma as result of electron SCX-mode kinetic damping:

$$L_{kin} \approx \tau\omega / \left(\upsilon_{Te} k_2^2\right), \tag{4.72}$$

where $\tau \approx 0.07$ and $\tau \approx 0.6$ in the limiting cases of thick and thin dielectric layer, respectively. The discharge length in the regime of low pressure of the working gas can be increased by applying a stronger external magnetic field and/or by increasing the wavelength. However, both possibilities are very restricted due to the necessity to keep the resonance condition between the wave frequency and the generator frequency. One can see from (4.72) that there also exists some possibility to increase $L_{kin}$ by making the protective dielectric coating thinner. These conclusions correspond to the known experimental results [68, 71] for plasma discharges, which are sustained by surface waves. The results of numerical studies of the discharge length are presented in Fig. 4.16. The pressure is assumed to be low, $p \approx 10$ mTorr, the thickness of the protective dielectric layer is $a = 1$ cm, and the generator frequency is $f = 2.45$ GHz. The volume of the plasma discharge is determined by the depth $\lambda_\perp$ of electron SCX-mode penetration into the plasma and the discharge length $L$.

The depth $\lambda_\perp$ was shown in [34] to be approximately equal to the SCX-mode wavelength, which in turn weakly depends on both the plasma particle density and the properties of the dielectric layer, which separates the plasma from the waveguide metal wall. The magnitude of both parameters, $\lambda_\perp$ and $L$, is determined mainly by the magnitude of the external magnetic field $B_0$. An increasing $B_0$ results in an increase of both $\lambda_\perp$ (under fixed magnitude of the working frequency $\omega$) and discharge length $L$, which means in a larger volume of the produced plasma.

What are the dimensions of a uniform plasma, which can be sustained by electron SCX-modes in a gas discharge with dielectrically coated plane parallel electrodes? If

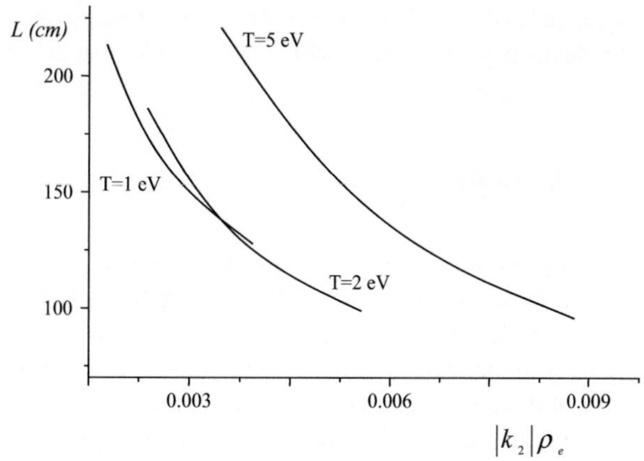


**Fig. 4.16** Length of plasma discharge sustained by electron SCX-modes versus wavenumber for different electron temperatures

the working frequency of the magnetron generator is $f = 2.45$ GHz, the external static magnetic field should be $B_0 = 500$ G to allow electron SCX-modes to be excited. Let the dielectric coating be thin and the plasma particle density and temperature be $n_{pl} \approx 10^{12}$ cm$^{-3}$ and $T \approx 1$ eV, respectively. In this case the relation between electron plasma and cyclotron frequencies is $\Omega_e^2 \approx 67 \cdot \omega_e^2$, and the average Larmor radius is $\rho_e \approx 0.006$ cm. For these conditions, the length of the discharge sustained by electron SCX-modes calculated according to (4.72) is $L \approx 0.07 \cdot k_2^{-2} \rho_e^{-1}$ cm. Thus, for a SCX-mode with the wavelength 3.7 cm the discharge length is approximately 70 cm, and the transverse size is about ~1 cm. If the wavelength is twice larger, the discharge length increases by four times. For the chosen parameters of the discharge plasma, $Q_{kin}/Q_{oh} \sim 7$ if the dielectric coating is thick, and $Q_{kin}/Q_{oh} \sim N_e^2$ if it is thin. This makes it possible to improve the conclusions, which were made in [76–78] about the ineffectiveness of collisionless mechanisms of energy transfer in microwave discharges: these conclusions are not valid for discharges sustained by surface type waves.

Thus, in this section, the electrodynamic model of plasma sources, which are sustained by electron SCX-mode propagation, is theoretically studied with account for Ohmic and kinetic channels of wave energy transfer to the plasmas. Average values of the electron SCX-mode energy flow and the absorbed wave energy are calculated in the case of high and low pressure of the working gas. The dimensions of the uniform plasma volume, which can be sustained by electron SCX-modes, are determined for different discharge conditions. The spatial distribution of the SCX-mode fields is investigated. The dependencies of both the depth of SCX-mode penetration into the plasma and the discharge length on the discharge structure parameters are studied.

The gas discharge parameters are shown to weakly depend on the properties of the dielectric coating of the chamber electrodes, but they can be controlled by means of the external static magnetic field. Increasing value of $B_0$ results in an increase of the discharge length and volume of the produced plasma. The discharge length is larger in the case of thin dielectric coating, and it increases with increasing wavelength of the electron SCX-modes, which sustain the discharge.

## 4.7 Conclusions

The dispersion properties of surface electromagnetic waves, which propagate in plasma filled metal magnetoactive waveguide structures at frequencies close to harmonics of the electron cyclotron frequency, are studied in the framework of kinetic theory with account for the possible presence of a protective dielectric coating on the walls of the metal waveguide which separates the plasma from the metal. An external static magnetic field is parallel to the plasma interface, and the surface mode, which is used to be called as electron surface cyclotron X-mode (SCX-mode), propagates along the interface perpendicular to the external magnetic field. Damping of these waves caused by plasma particle collisions both with each other and with the plasma boundary is studied.

If there is no dielectric coating and the plasma has direct contact with the metal wall, the propagation of SCX-modes is unidirectional and the direction coincides with that of electron gyration nearby the plasma–metal interface. The presence of the protective dielectric coating results in removing this phenomenon of unidirectionality and the surface electron cyclotron waves can propagate in both directions across the external magnetic field.

The additional mode caused by the presence of the dielectric layer has a longer wavelength than the main mode. Increase in the magnitude of the dielectric constant as well as decreasing layer thickness result in slowing down the phase velocity of these modes. The properties of these modes are most sensitive to the value of the external magnetic field. Decrease of the plasma particle density in the transient layer at the plasma interface with the dielectric coating results in decreasing wave eigenfrequencies.

Parametric excitation of surface electron cyclotron waves with both monochromatic and non-monochromatic pumping RF fields is investigated. The threshold for the development of parametric instabilities, over which the growth rate is larger than the damping rate, is determined. The growth rate of parametric instabilities weakly depends on the plasma particle density and the frequency of the pumping RF field—they increase with decreasing plasma layer thickness and increasing pumping field amplitude. Non-monochromatic pumping RF fields significantly change the scenario of the development of parametric instabilities. In this case, the dependence of the growth rate on the RF field frequency and amplitude is non-monotonous. In the case of a two-frequency pumping RF field the growth rate reaches its maximum under the following conditions. The frequency of one of the pumping RF field components should be close to the corresponding harmonic of the electron cyclotron frequency. Then the amplitude of the second (non-resonant) pumping RF-field component can be used for effective variation of the growth rate and for controlling instability development.

An electrodynamic model of plasma sources sustained by surface electron cyclotron X-mode waves is proposed. The parameters of the gas discharge weakly depend on the properties of the protective dielectric coating of the discharge chamber. However, they are sensitive to the value of the external stationary magnetic field. Increase in the magnitude of the external magnetic field results in larger discharge length and volume of the produced plasma. The dimensions of the uniform plasma space, which can be sustained in the gas discharge by the electron SCX-modes, are determined.

## References

1. Bluyssen, H., Maan, J. C., Wyder, P., et al. (1975). Cyclotron resonance in an InAs-GaSb superlattice. *Solid State Communications, 31*(1), 35–38.
2. Maan, J. C., Guldner, Y., Vieren, J. P., et al. (1981). Three dimensional character of semimetallic InAs-GaSb superlattices. *Solid State Communications, 39*(5), 683–686.
3. Guldner, Y., Vieren, J. P., Voisin, P., et al. (1982). Observation of double cyclotron resonance and interband transitions in InAs-GaSb multi-heterojunctions. *Solid State Communications, 41*(6), 755–758.
4. Bass, F. G., Bulgakov, A. A., & Tetervov, A. P. (1989). *High-frequency properties of semiconductors with superlattices*. Moscow: Nauka. (in Russian).
5. Brazis, R. S. (1981). Active and nonlinear interactions at the excitation of plasma-like polaritons in semiconductors. *Litovskiy Fizicheskiy Sbornik, 21*(4), 73–117. (in Russian).
6. Stix, T. H. (1990). Waves in plasmas: Highlight from the past and present. *Physics of Fluids B, 2*(8), 1729–1743.
7. Kaner, E. A. (1958). Theory of cyclotron resonance. *Journal of Experimental and Theoretical Physics, 6*(6), 1135–1138.
8. Kaner, E. A., & Skobov, V. G. (1967). Electromagnetic waves in metals in a magnetic field. *Soviet Physics Uspekhi, 9*(4), 480–503.
9. Vedenov, A. A. (1965). Solid state plasma. *Physics-Uspekhi, 7*(6), 809–822.
10. Lifshits, I. M., Azbel, M., & Kaganov, M. I. (1973). *Electron theory of metals*. New York: Consultants Bureau.
11. Steele, M., & Vural, B. (1969). *Wave interactions in solid state plasmas*. New York: McGraw Hill.
12. Vladimirov, V. V., Volkov, A. F., & Meylikhov, Ye. Z. (1979). *Plasma of semiconductors*. Moscow: Atomizdat. (in Russian).
13. Platzman, P. M., & Wolff, P. A. (1973). *Waves and interactions in solid state plasmas*. New York: Academic Press.
14. Beletski, N. N., Bulgakov, A. A., Khankina, S. I., & Iakovenko, V. M. (1984). *Plasma instabilities and non-linear phenomena in semi-conductors*. Kyiv: Naukova Dumka. (in Russian).
15. Dmitruk, N. L., Litovchenko, V. G., & Strizhevskiy, V. L. (1989). *Surface polaritons in semiconductors and dielectrics*. Kiev: Naukova Dumka. (in Russian).
16. Damon, R. W., & Eshbach, J. R. (1961). Magnetostatic modes of a ferromagnetic slab. *Journal of Physics and Chemistry of Solids, 19*(3/4), 308–320.
17. Kaner, E. A., & Makarov, N. M. (1970). Theory of surface electromagnetic waves in metals located in a weak magnetic fieid. *Soviet Physics JETP, 31*(6), 1063–1070.
18. Silin, V. P., & Tolkachev, O. M. (1978). Cyclotron resonance on the skipping orbits in an electron liquid. *Soviet Physics JETP, 47*(6), 1113–1121.
19. Silin, V. P., & Tolkachev, O. M. (1980). Quantum theory of cyclotron resonance on sliding orbits near the frequencies of transitions between the levels of surface electrons in the degenerate electronic fluid. *Fizika tverdogo tela, 22*(2), 374–382. (in Russian).

20. Man'kov, Y. I. (1980). Longitudinal electromagnetic waves in a metal in inclined magnetic field nearby the cyclotron resonance. *Fizika tverdogo tela, 22*(2), 408–412. (in Russian).
21. Walsh, W. M., Platzman, J., & Platzman, P. M. (1965). Excitation of plasma waves near cyclotron resonance in potassium. *Physical Review Letters, 15*(20), 784–786.
22. Platzman, P. M., Walsh, W. M., & Foo, E.-N. (1968). Fermi-liquid effects on high-frequency wave propagation in simple metals. *Physical Review, 172*(3), 689–698.
23. Foo, E.-N. (1969). Dispersion of ordinary and extraordinary high-frequency waves in metals. *Physical Review, 182*(3), 674–678.
24. Kondratenko, A. N. (1979). *Penetration of a field into plasmas*. Moscow: Atomizdat. (in Russian).
25. Akhiezer, A. I., Akhiezer, I. A., Polovin, R. V., Sitenko, A. G., & Stepanov, K. N. (1975). *Plasma electrodynamics*. Oxford: Pergamon Press.
26. Lominadze, D. G. (1981). *Cyclotron waves in plasma*. Oxford: Pergamon Press.
27. Aleksandrov, A. F., Bogdankevich, L. S., & Rukhadze, A. A. (1984). *Principles of plasma electrodynamics*. Berlin: Springer.
28. Kondratenko, A. N. (1976). *Plasma waveguides*. Moscow: Atomizdat. (in Russian).
29. Krall, N. A., & Trivelpiece, A. W. (1986). *Principles of plasma physics*. San Francisco: San Francisco Press.
30. Brambilla, M. (1998). *Kinetic theory of plasma waves. Homogeneous plasmas*. Oxford: Clarendon Press.
31. Girka, V. O., Kondratenko, A. M., & Pavlenko, I. V. (1993). Surface cyclotron waves in the planar metal waveguide with plasma filling. In *Proceedings and Contributed Papers of International Conference "Physics in Ukraine"* (pp. 108–110), Ukraine.
32. Girka, V. O., & Pavlenko, I. V. (1993). To the kinetic theory of surface waves at the plasma-metal interface. *Ukrainain Journal of Physics, 38*(4), 529–533. (in Russian).
33. Kondratenko, A. N. (1985). *Surface and bulk waves in bounded plasmas*. Moscow: Energoatomizdat. (in Russian).
34. Ginzburg, V. L., & Rukhadze, A. A. (1972). *Waves in magnetoplasma*. Heidelberg: Springer.
35. Sitenko, O. G., & Mal'nev, V. M. (1994). *Principles of plasma theory*. Kyiv: Naukova Dumka. (in Ukrainian).
36. Azarenkov, N. A., Girka, V. O., Kondratenko, A. M., et al. (1997). Surface waves on the harmonics of the electron cyclotron frequency propagating along a plasma–metal interface. *Plasma Physics and Controlled Fusion, 39*, 375–388.
37. Girka, V. A., Kondratenko, A. N., & Podobinski, V. O. (1981). Extraordinary surface cyclotron waves in semi-bonded plasma under the conditions of weak spatial dispersion. *Ukrainain Journal of Physics, 26*(4), 681–683. (in Russian).
38. Azarenkov, N. A., & Ostrikov, K. N. (1999). Surface magnetoplasma waves at the interface between a plasma-like medium and a metal in a Voigt geometry. *Physics Reports, 308*, 333–428.
39. Azarenkov, N. A., Girka, V. O., & Sporov, A. E. (1997). Surface cyclotron waves in a nonuniform plasma filled metal waveguide with dielectrical coating. *Physica Scripta, 55*, 339–344.
40. Laqua, H. P., Erckmann, V., Hartfuβ, H. J., et. al. (1996). Resonant and non-resonant electron cyclotron heating at densities above the plasma cut-off by O–X–B mode conversion at W7-AS. In *23rd European Physical Society Conference on Controlled Fusion and Plasma Physics* (pp. 847–850), Ukraine.
41. Silin, V. P. (1973). *Parametric effect of high-intensity irradiation on plasmas*. Moscow: Nauka. (in Russian).
42. Galeev, A. A., & Sagdeev, R. Z. (1973). Parametric phenomena in a plasma. *Nuclear Fusion, 13*(4), 603–621.
43. Aliev, Y. M., Silin, V. P., & Watson, C. (1966). Parametric resonance in a plasma in a magnetic field. *Soviet Physics JETP, 23*(4), 626–632.
44. Silin, V. P. (1967). Ion cyclotron instability of plasma in strong high frequency field. *Soviet Physics Technical Physics, 37*(5), 991–993. (in Russian).
45. Tzoar, N. (1969). Parametric excitation in plasma in a magnetic field. *Physical Review, 178*(1), 356–362.

46. Aliev, Y. M., & Zunder, D. (1970). Parametric excitation of upper and lower hybrid resonances. *Soviet Physics JETP, 30*(4), 718–720.
47. Andreev, N. E., & Kiriy, A. Y. (1971). To the theory of plasma instability in HF electric and stationary magnetic fields. *Soviet Physics Technical Physics, 41*(6), 1080–1087. (in Russian).
48. Andreev, N. E. (1971). Parametric instability of a plasma in a constant magnetic field and a weak high-frequency electric field. *Radiophysics and Quantum Electronics, 14*(8), 909–914.
49. Aliev, Y. M., & Silin, V. P. (1973). Parametric effect on a plasma of high intensity irradiation near electron cyclotron frequencies. *Soviet Physics Technical Physics, 17*(1), 1752–1767.
50. Kitsenko, A. B., Panchenko, V. I., & Stepanov, K. N. (1973). Electron-acoustic and ion cyclotron parametric instabilities of a plasma in AC electric field. *Soviet Physics Technical Physics, 43*(7), 1426–1444. (in Russian).
51. Kitsenko, A. B., Lominadze, D. G., & Stepanov, K. N. (1974). The parametric excitation of the electron cyclotron oscillations of a plasma located in an alternating electric field. *Soviet Physics JETP, 39*(2), 294–298.
52. Stepanov, K. N. (1996). Nonlinear parametric phenomena in plasma during radio frequency heating in the ion cyclotron frequency range. *Plasma Physics and Controlled Fusion, 38,* A13–A29.
53. Longinov, A. V., & Stepanov, K. N. (1992). Radio-frequency plasma heating in tokamaks in the ion-cyclotron frequency range. In A. G. Litvak (Ed.), *High-frequency plasma heating* (pp. 93–237). New York: AIP.
54. Becoulet, A. (1996). Heating and current drive regimes in the ion cyclotron range of frequency. *Plasma Physics and Controlled Fusion, 38*(12A), 1–12.
55. Grigor'eva, L. I., Smerdov, B. I., Stepanov, K. N., et al. (1970). Plasma instability in a strong alternating electric field. *Soviet Physics JETP, 31*(1), 26–28.
56. Azarenkov, N. A., Kondratenko, A. N., & Ostrikov, K. N. (1990). Parametric excitation of surface waves at the plasma-metal interface. *Soviet Physics Technical Physics, 69*(1), 31–36. (in Russian).
57. Aliev, Y. M., & Ferlengi, E. (1970). Parametric excitation of surface oscillations of a plasma by an external high frequency field. *Soviet Physics JETP, 30*(5), 877–879.
58. Aliev, Y. M., Gradov, O. M., & Kirii, A. Y. (1973). Kinetic theory of parametric excitation of surface waves in a semi-finite plasma. *Soviet Physics JETP, 36*(1), 59–63.
59. Lovetski, E. E., & Starodub, A. N. (1974). Surface oscillations of magnetoactive plasma in high frequency field. *Soviet Physics Technical Physics, 44*(3), 508–513. (in Russian).
60. Dragila, R., & Vucovic, S. (1988). Excitation of surface waves by an electromagnetic wave packet. *Physical Review Letters, 61*(24), 2759–2761.
61. Aliev, Y. M., & Gradov, O. M. (1972). Parametric excitation of surface waves in nonuniform magnetized plasma. *Soviet Physics Technical Physics, 42*(11), 2447–2448. (in Russian).
62. Girka, V. A., & Lapshin, V. I. (1987). Parametric surface-wave excitation at the second harmonics of ion and electron cyclotron frequencies. *Radiophysics and Quantum Electronics, 30*(6), 544–547.
63. Girka, V. O., Girka, I. O., Kondratenko, A. M., et al. (1996). Surface electron cyclotron waves in the metallic waveguide structures with two component filling. *Contributions to Plasma Physics, 36*(6), 679–686.
64. Korzh, A. F., & Stepanov, K. N. (1988). Parametric instability of plasma in electric field of two pumping waves. *Fizika Plazmy, 14*(6), 698–705. (in Russian).
65. Puzirkov, S. Y., Girka, V. O., & Lapshin, V. I. (1998). Parametric excitation of the surface cyclotron waves by nonmonochromatic electric field. In *Proceedings of the 1998 International Conference on Mathematical Methods in Electromagnet. Theory* (pp. 697–699), Ukraine.
66. Girka, V. O., Lapshin, V. I., & Puzirkov, S. Y. (1999). Parametric instability of surface type X-modes immersed into a nonmonochromatic pumping electric field. *Contributions to Plasma Physics, 39*(6), 487–494.
67. Moisan, M., & Zakrzewski, Z. (1986). Plasmas sustained by surface waves at microwave and RF frequencies: Experimental investigation and applications. In J. M. Proud & L. H. Luessen (Eds.), *Radiative processes in discharge plasmas. NATO ASI, Series B* (Vol. 149, pp. 381–430).

68. Zhelyazkov, I., Atanasov, V., & Benova, E. (1986). In S. Vucovic (Ed.), *Surface waves in plasmas and solids*. Singapore: World Scientific.
69. Ferreira, C. M., & Moisan, M. (1993). Microwave discharges: Fundamentals and applications. In *NATO Advanced Study Institute, Series B: Physics* (Vol. 302, pp. 187–544).
70. Zhelyazkov, I., & Atanasov, V. (1995). Axial structure of low-pressure high-frequency discharges sustained by travelling electromagnetic surface waves. *Physics Reports, 255,* 79–201.
71. Margot, J., & Moisan, M. (1991). Electromagnetic surface waves for a new approach to the investigation of plasmas produced at electron cyclotron resonance. *Journal of Physics D: Applied Physics, 24*(9), 1765–1788.
72. Margot, J., & Moisan, M. (1992). Surface wave sustained plasmas in static magnetic fields for study of ECR discharge mechanisms. In M. Moisan & J. Pelletier (Eds.), *Microwave exited plasmas* (pp. 229–48). Amsterdam: Elsevier.
73. Girka, V. O. (1998). Theoretical model of a stationary low-pressure plasma source based on surface type X-mode. *Physica Scripta, 58*(4), 387–391.
74. Kortshagen, U., Schluter, H., & Shivarova, A. (1991). Determination of electron energy distribution functions in surface waves produced plasma: I. Modelling. *Journal of Physics D: Applied Physics, 24,* 6063–6078.
75. Shaing, K. C. (1996). Electron heating in inductively coupled discharges. *Physics of Plasmas, 3*(9), 3300–3303.
76. Turner, M. M. (1993). Collisionless electron heating in an inductively coupled discharge. *Physical Review Letters, 71*(12), 1844–1847.
77. Turner, M. M. (1995). Pressure heating of electrons in capacitively coupled RF discharges. *Physical Review Letters, 75*(7), 1312–1315.
78. Kortsgahen, U., Busch, C., & Tsendin, L. D. (1996). On simplifying approaches to the solution of the Boltzmann equation in spatially inhomogeneous plasma. *Plasma Source Science and Technology, 5,* 1–17.
79. Kolobov, V. I., & Godyak, V. A. (1995). Nonlocal electron kinetics in collisional gas discharge plasmas. *IEEE Transactions on Plasma Science, 23*(33), 503–531.
80. Tatarova, E., Dias, F. M., Ferreira, C. M., et al. (1997). Self-consistent kinetic model of a surface wave sustained discharge in nitrogen. *Journal of Physics D: Applied Physics, 30*(19), 2663–2676.
81. Chen, F. F. (1995). Industrial application of low-temperature plasma physics. *Physics of Plasmas, 2*(6), 2164–2175.
82. Chen, F. F. (1996). Physics of helicon discharges. *Physics of Plasmas, 3*(5), 1783–1790.

# Chapter 5
# Surface Electron Cyclotron O-Mode Waves

Studying surface type waves is of great interest both for plasma physics [1–9] and physics of solids [7, 10–24]. This is associated, first of all, with a wide spectrum of technological applications. Main applications are e.g. RF driven gas discharges, which are produced and sustained by surface waves with the purpose of processing large-area semiconductors, pumping of active laser media, etc. [25–33]. Ion and electron cyclotron resonance waves are known to be widely applied for additional plasma heating in magnetic thermonuclear fusion devices (see, e.g., [34–38] and references therein). Just bulk cyclotron waves are certainly used for these purposes. Their properties are already well studied [1, 39–51] unlike those of surface waves. The bulk cyclotron waves are known to be divided into three types with respect of their polarization, namely: ordinary and extraordinary polarized, as well as longitudinal cyclotron waves, the Bernstein waves.

In the previous chapters, the results of studying the properties of surface TM- and X-modes at harmonics of the electron cyclotron frequency are presented. This explains the choice of the subject to be studied in the present chapter. Namely, this chapter complements the research carried out in the previous chapters by investigating the possibility of surface electron cyclotron O-mode wave propagation. The problem is studied theoretically in the kinetic non-relativistic approach for the model of a planar metal waveguide which is filled by plasma separated from the waveguide metal wall by a protective dielectric layer. The external stationary magnetic field is assumed to be oriented parallel to the plasma interface. It should be noted that the problem of specific damping of these surface waves, that is caused by the relativistic Doppler effect and which can take place in the case when the eigenfrequency is very close to the corresponding cyclotron resonance, remains out of scope of this monograph. Such damping was studied for the bulk cyclotron waves (see [39] and references therein) for the case when the wave frequency was lower than the corresponding cyclotron resonance.



V. Girka et al., *Surface Electron Cyclotron Waves in Plasmas*,
Springer Series on Atomic, Optical, and Plasma Physics 107,
https://doi.org/10.1007/978-3-030-17115-5_5

## 5.1 Surface Electron Cyclotron O-Mode Propagation in Waveguide Structures Semibounded Plasma–Dielectric–Metal

The geometry of the problem is as follows (see Fig. 5.1). A semibounded uniform plasma occupies the half-space $x \geq 0$ and has an interface with a dielectric layer (this also can be vacuum) with thickness $a_{\rm d}$ and dielectric constant $\varepsilon_{\rm d}$. The dielectric layer occupies the region $-a_{\rm d} \leq x \leq 0$. The waveguide metal wall is placed in the region $x < -a_{\rm d}$. The external stationary magnetic field $\vec{B}_0$ is parallel to the plasma interface and is oriented along the $\vec{z}$ axis. The transverse dimensions of the plasma are assumed to be much larger than the depth of wave penetration into the plasma—just in this case the model of semibounded plasma is valid.

The electromagnetic fields of surface electron cyclotron wave modes with ordinary polarization (O-modes) are described by the Maxwell equations. The dependence of the fields on coordinates and time is chosen in the following form: $E, H \propto f(x) \times \exp(ik_2 y - i\omega t)$, where $\omega$ is the angular wave frequency and $k_2$ is the component of the wave vector along the $\vec{y}$ axis. The space is assumed to be isotropic along the $\vec{z}$ axis. In this case there is no dependence on this coordinate, which means $\partial/\partial z \equiv 0$. The spatial dispersion of the plasma is assumed to be weak, so that the following inequality is valid: $k_\perp v_{T\alpha} \ll \omega - s|\omega_\alpha|$, here $k_\perp^2 = k_1^2 + k_2^2$, where $k_1$ is the component of the wave vector along the $\vec{x}$ axis, $v_{T\alpha}$ and $\omega_\alpha$ are thermal velocity and cyclotron frequency of plasma particles ($\alpha = e$ for electrons and $\alpha = i$ for ions), respectively, and $s$ is an integer, the number of cyclotron harmonic. The present investigation is restricted to the case of waves propagating at the first electron cyclotron harmonic. This restriction is explained by the fact that the dispersion relation cannot be derived in a general case (for an arbitrary value of $s$) similar to the cases studied in the previous Chaps. 3 and 4.

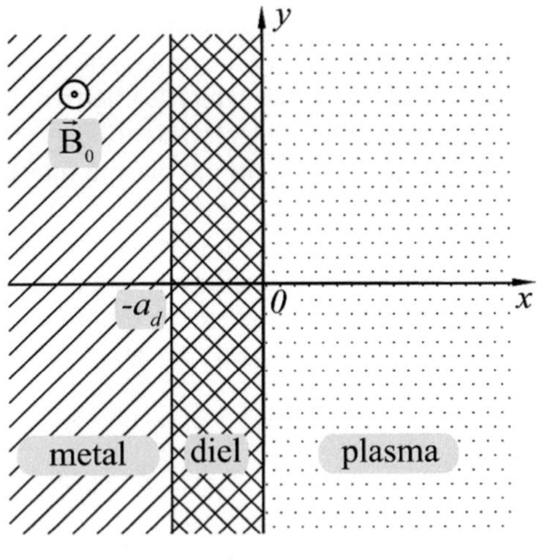


**Fig. 5.1** Schematic of the problem “semibounded plasma–dielectric–metal”

The motion of electrons is described by the kinetic Vlasov–Boltzmann equation where the integral of collisions is taken into account in the approach of an effective collision frequency. The equilibrium state of the plasma is described by a Maxwellian distribution function. The anti-Hermitian tensor $A_{ij}$ describes the contribution to plasma conductivity of those electrons, which collide with the plasma interface and thereby change their phase of Larmor gyration in the external static magnetic field in different way as compared to the electrons which do not collide with the interface. The tensor $A_{ij}$ has a complicated integral form. In the limiting case of weak plasma spatial dispersion for the model of electron diffusive reflection from the plasma interface, damping of electron surface cyclotron O-modes (SCO-modes) was studied in [52]. There, the longitudinal (with respect to the plasma interface) component of the wave vector $\vec{k}_\perp$ was not taken into account during solving the kinetic equation. This assumption certainly allowed to simplify the finding of roots but reduced the calculation precision of the numerical coefficients in the expressions for the dispersion relation. In this case the approximate expression for the kernel of the conductivity tensor $A_{33}$ can be written as follows:

$$A_{33} = \sum_\alpha \frac{\upsilon_{T\alpha}\Omega_\alpha^2}{\pi^{3/2}c\omega^2}\left[1 - \frac{1+\exp(2i\pi\omega/\omega_\alpha)}{2-8(\omega/\omega_\alpha)^2}\right] \approx \frac{\upsilon_{Te}\Omega_e^2}{\pi^{3/2}c\omega^2}, \tag{5.1}$$

where $\Omega_e$ is the electron plasma frequency. If the inequality $\upsilon_{Te} \ll c$ is true, the contribution of $A_{33}$ to the expression of the axial RF current can be neglected as compared with the diagonal element of the conductivity tensor of an infinite plasma: $|\sigma_{33}| \gg A_{33}$.

Thus, one can use the expression for the plasma permittivity tensor $\varepsilon_{ik}$ obtained for the model of an infinite magnetoactive plasma (see, e.g., [39, 40]) when studying the dispersion properties of slow SCO-modes. In this case the Fourier coefficients $j_3(k_\perp)$ and $E_3(k_\perp)$ of the axial current density and the wave axial electric field are linked with each other by the following algebraic correlation:

$$j_3(k_\perp) = \sigma_{33}(k_\perp)E_3(k_\perp). \tag{5.2}$$

In (5.2), $\sigma_{33}$ is the component of the permittivity tensor of an infinite magneto-active plasma [39, 40]:

$$\sigma_{33} = \frac{i\omega}{2\pi}\sum_\alpha \sum_{s=0}^{\infty}{}^{*} \frac{\Omega_\alpha^2 I_s(k_\perp^2\rho_\alpha^2)}{\omega^2 - s^2\omega_\alpha^2}, \tag{5.3}$$

where $I_n(x)$ is a modified Bessel function [53], $\rho_\alpha$ is the Larmor radius of plasma particle specie $\alpha$, and the superscript * nearby the symbol of summation with respect to the number of cyclotron harmonic $s$ means that the term with the number $s = 0$ should be multiplied by 1/2. Based on the initial assumption of weak plasma spatial dispersion, which is mathematically expressed by the inequality $k_\perp^2\rho_\alpha^2 \ll 1$, one can use the asymptotic expression $I_n(x) \approx (x/2)^n/n!$ for the modified Bessel function in (5.3).

Since it is foreseen to study electron SCO-modes at the first harmonic of the electron cyclotron frequency, one can keep only two terms in the sum (5.3) over the numbers of cyclotron harmonics [39, 40]. The first term describes the component $\sigma_3$ of the plasma electric conductivity in hydrodynamic approach (this term has the number of summation $s = 0$). The second term (with $s = 1$) takes into account the finite dimensions of the Larmor radius of gyration of the charged plasma particles in the external static magnetic field, and makes it possible to describe electromagnetic waves at the first harmonic of the electron cyclotron frequency correctly.

Despite just electron SCO-modes are under the consideration, one has to evaluate the contribution of the ion terms to the value of $\sigma_{33}$ for $\omega \sim |\omega_e|$. The magnitude of the ion temperature is assumed to be of the order or even lower than that of electrons. Keeping in mind the following inequalities: $\Omega_e^2 \gg \Omega_i^2$, $k_\perp^2\ \rho_e^2 \ll k_\perp^2\ \rho_i^2 \ll 1$, and $\omega^2 - \omega_e^2 \ll \omega^2 - \omega_i^2 \sim \omega^2$, one can write the component $\sigma_{33}$ of the plasma conductivity tensor in the following approximate form in this frequency range:

$$\sigma_{33} \approx \frac{i\Omega_e^2}{4\pi\omega} + \frac{i\omega\,\Omega_e^2 k_\perp^2 \rho_e^2}{4\left(\omega^2 - \omega_e^2\right)}. \tag{5.4}$$

For studying ion SCO-modes, which propagate at the ion cyclotron frequency, $\omega \sim \omega_i$, the component $\sigma_{33}$ has the corresponding following approximate form:

$$\sigma_{33} \approx \frac{i\Omega_e^2}{4\pi\omega} + \frac{i\omega\,\Omega_i^2 k_\perp^2 \rho_i^2}{4\left(\omega^2 - \omega_i^2\right)}. \tag{5.5}$$

Comparison of the expression (5.4) with (5.5) confirms that one can neglect the ion contribution to $\sigma_{33}$ when studying the properties of electron SCO-modes.

Under these conditions the set of Maxwell equations can be separated into two independent sets. One of these sets describes the ordinarily polarized wave with the field components $H_x$, $H_y$, $E_z$ (its electric field component is parallel to the external static magnetic field). Unlike the electron SCX-modes studied in the previous chapter, the fields of SCO-modes have perfect symmetry, namely, $H_x$ and $E_z$ are even functions of the transverse coordinate $x$, and $H_y$ is an odd function: $H_y(+x) = -H_y(-x)$. Using the Fourier method makes it possible to get the following set of equations for the Fourier coefficients of electron SCO-mode fields from the set of Maxwell equations:

$$\begin{aligned} kH_1 &= k_2 E_3, \\ kH_2 &= -k_1 E_3, \\ ik_1 H_2 - \frac{H_y(0)}{\pi} &= ik_2 H_1 - ikE_3 + \frac{4\pi}{c} j_3(k_1). \end{aligned} \tag{5.6}$$

In (5.6), $H_y(0)$ is the value of the tangential component of the wave magnetic field at the plasma interface, and $kc = \omega$.

Solving the (5.2) and (5.6) one derives the following algebraic equation for $E_3(k_1)$:

$$\left(k_2^2 + k_1^2 - k^2\varepsilon_{33}\right)E_3(k_1) = \frac{ik\,H_y(0)}{\pi} + iA_{33}E_z(0). \tag{5.7}$$

In (5.7) $\varepsilon_{33}(k_1)$ is the component of the magnetoactive plasma permittivity tensor calculated in the approach of an infinite plasma with a weak spatial dispersion under the validity of the inequality $h_\alpha \gg v_{T\alpha}^2 c^{-2}$, which is called as non-relativistic approach [1, 39, 40],

$$\varepsilon_{33} \approx \varepsilon_3 - \sum_\alpha \sum_{s=-\infty}^{+\infty} \frac{\Omega_\alpha^2\left(k_1^2\rho_\alpha^2\right)^s}{\omega^2 s!h_\alpha(s)}, \tag{5.8}$$

where $\varepsilon_3 = 1 - \left(\Omega_e^2 + \Omega_i^2\right)/\omega^2 \approx 1 - \Omega_e^2/\omega^2$ is the diagonal element of the plasma permittivity tensor calculated in hydrodynamic approach, and the parameter $h_\alpha(s)$ characterizes the deviation of the wave frequency from the corresponding harmonic of the cyclotron frequency: $h_\alpha = 1 - s\omega_\alpha/\omega$. The other notations are the same as in the previous two chapters. The analysis of the expression (5.8) for $\varepsilon_{33}(k_1)$ shows that in this case, there is no dependence on the sign of $k_1$, unlike in the case of electron SCX-modes studied in Chap. 4. The dispersion properties of electron SCO-modes significantly depend on the component $\varepsilon_{33}(k_1)$ of the plasma dielectric permittivity tensor. In this respect SCO-modes are similar to the corresponding bulk modes. The diagonal element $\varepsilon_{33}$ of the plasma permittivity tensor contains the wavenumber of SCO-modes in even symmetry. Therefore, the sign of $k_1$ does not influence the dispersion properties of these modes.

Inverse Fourier transformation of (5.7) makes it possible to derive the integral expressions for the SCO-mode fields in the region occupied by the plasma. The expression for the electric component of electron SCO-modes is as follows:

$$E_z(x) = \int_{-\infty}^{+\infty} \frac{ik}{\pi\,\Delta(k_1)}\left[H_y(0) + \pi A_{33}E_z(0)\right]\exp(ik_1x)\,\mathrm{d}k_1, \tag{5.9}$$

where $\Delta(k_1) = k_1^2 + k_2^2 - k^2\varepsilon_{33}(k_1)$. This improper integral (which is to be calculated over the transverse wavenumber) should be replaced by a contour integral in the upper complex half-plane of $k_1$. Then it can be calculated by the theory of residues. The cumulative function of this integral has $s$ poles in the upper half-plane. Their position is determined by the roots of the equation $\Delta(k_1) = 0$. It is a pity but unlike in the case studied in Chap. 3, the integral cannot be calculated in general form in the case of an arbitrary number of the cyclotron harmonic. In the case of electron SCO-modes at the first harmonic of the electron cyclotron frequency, the equation which determines the poles of the cumulative function in (5.9) has the following form:

$$k_1^2\left[1+\frac{\Omega_e^2\rho_e^2}{2c^2h_e(1)}\right]+k_2^2-k^2\varepsilon_3\approx 0. \tag{5.10}$$

To find the SCO-mode electric field in the plasma region in this case of $s = 1$ one has to calculate only one residue for the root of (5.10) in the upper complex half-plane which has the following form:

$$k_1 = i\sqrt{\frac{(k_2^2\delta^2+1)2h_e(1)}{\rho_e^2+2\delta^2h_e(1)}}, \tag{5.11}$$

where $\delta = c/\Omega_e$ is the skin-depth. Then one can derive the following expression for the electron SCO-mode electric field from (5.9):

$$E_z(x) = -2k\left\{\frac{\partial\Delta}{\partial k_1}\right\}^{-1}\left[H_y(0)+\pi A_{33}E_z(0)\right], \tag{5.12}$$

where the derivative of $\Delta(k_1)$ should be calculated for the $k_1$ value given by (5.11).

It follows from the analysis of (5.9) that even in the more or less simple case $s = 2$, the equation $\Delta(k_1) = 0$ gets a very cumbersome form. That is why it seems to be impossible to find a simple analytic expression for the dependence $\omega = \omega(k_2)$ for higher numbers of harmonics of the electron cyclotron frequency.

To derive the dispersion relation for electron SCO-modes, one has to find the expressions for the tangential components of the wave fields from the set of Maxwell equations also in the region of the dielectric coating and apply the following boundary conditions.

(1) The component $E_z(x)$ of the electric wave field in the dielectric layer should be equal to zero at the dielectric–metal interface $x = -a_{\mathrm{d}}$.
(2) The component $E_z(x)$ of the electric wave field should be continuous at the plasma–dielectric interface $x = 0$.
(3) The tangential component of the magnetic wave field is also continuous at the plasma–dielectric interface.

One can derive the following equations from the set of Maxwell equations for electron SCO-mode fields in the space occupied by the dielectric:

$$\frac{\mathrm{d}H_y}{\mathrm{d}x} = \frac{i}{k}(k_2^2-k^2)E_z(x);\quad ikH_y = -\frac{\mathrm{d}E_z}{\mathrm{d}x}. \tag{5.13}$$

Second order differential equations for the tangential components of these surface wave fields in vacuum can be derived from the set of equations (5.13). If the waves are sufficiently slow such that the inequality $c^2k_2^2 \gg \omega^2$ is valid, the impedance has the following form at the dielectric interface $x = 0$:

$$\frac{E_z(0)}{H_y(0)} \approx \frac{ik}{k_2\tanh(a_{\mathrm{d}}k_2)}. \tag{5.14}$$

Keeping in mind the result of calculating the integral (5.9) with the help of the theory of residues one can derive the following dispersion relation for SCO-modes at the first harmonic of the electron cyclotron frequency:

$$k_2^2 - k^2\varepsilon_3\left[1 + \rho_e^2/4\delta^2 h_e(1)\right] = k_2^2 \coth^2(a_d k_2). \tag{5.15}$$

In (5.15) the terms, which are proportional to $A_{33}$ and determine wave damping, are omitted in studying the dispersion properties of SCO-modes. For the same reasons, the presence of the collision frequency in the component $\varepsilon_{33}$ of the plasma permittivity tensor is ignored.

In the limiting case of a thin dielectric layer ($a_d \gg \delta$) the eigenfrequency of electron SCO-modes is lower than the electron cyclotron frequency:

$$h_e(1) \approx -0.25\rho_e^2\left(1 + k_2^2\delta^2\right)\delta^{-2} < 0. \tag{5.16}$$

To meet the initial assumption of a non-relativistic approach the expression (5.16) should be used for description of SCO-modes in a plasma with sufficiently high plasma particle density, $\Omega_\alpha^2 \gg \omega_\alpha^2$, the wavelength should be relatively large, $k_2^2\delta^2 \ll 1$, and the plasma temperature should be low enough so that $\rho_e^2 \ll \delta^2$.

A more general expression for the electron SCO-mode eigenfrequency has the following form:

$$\left(\frac{\omega}{\omega_e}\right)^2 = \left(1 + \frac{\rho_e^2\delta_e^{-2}}{1 + 0.5k_2^{-2}\delta_e^{-2}\left(1 + \sqrt{1 + 4k_2^4\delta_e^4 \coth^2(a_d k_2)}\right)}\right)^{-1}. \tag{5.17}$$

The analysis of expression (5.17) makes it possible to conclude that the frequency of a SCO-mode at the fundamental harmonic of the electron cyclotron frequency is lower than $|\omega_e|$ [52]. The SCO-mode frequency shift $h_e(s)$ from the corresponding cyclotron resonance cannot have an infinitely small value since its decrease is associated with increasing wave damping rate caused by plasma particle collisions: a lower threshold for the $h_\alpha(s)$ value is given by the inequality $\nu \ll \omega h_\alpha(s)$. SCO-mode waves have reverse dispersion. Their group velocity decreases with decreasing thickness of the dielectric layer, which separates the plasma from the waveguide metal wall, and with decreasing plasma beta, $\beta_e = 4\pi\, n_e T_e/B_0^2$, that is the fraction of gas-kinetic pressure to that of the magnetic field.

Electron SCO-modes cannot propagate along the plasma–metal interface unlike SCX-modes. This is explained by the fact that the only electric component of SCO-mode fields is tangential to the plasma interface. To meet the boundary condition, which consists in vanishing tangential electric wave field $E_z(x)$ at the metal interface $x = 0$, the electric field of SCO-modes has to be equal to zero in the whole plasma volume in the case of the waveguide structure plasma–metal (with no dielectric layer).

The magnetic components of SCO-mode fields are proportional to $E_z(x)$. That is why they follow the electric component and are equal to zero in the waveguide structure plasma–metal. In general, a reduction of the thickness of the dielectric layer worsens the conditions of electron SCO-mode propagation.

The numerical analysis of the dispersion relation (5.15) makes it possible to get the graphical dependences of SCO-mode eigenfrequencies on the wavenumber. The product $k_2\rho_e$ is used as abscissa axis in Fig. 5.2. The numerical results confirm the analytical predictions. This follows from comparison of the dispersion curves of the electron SCO-modes presented in Fig. 5.2, which are obtained for different values of the plasma beta, $\beta_e$. Moreover, the numerical analysis carried out in the later Sect. 5.3 proves that the expression (5.17) satisfactorily describes also the case of a plasma with inhomogeneous profile of the plasma particle density. The depth of SCO-mode penetration into the plasma is larger than the electron Larmor radius.

The results presented here are calculated for the case of a Cartesian system of coordinates. With respect to studying the dispersion properties of electron SCO-modes, these results coincide with those presented in [9] where waves at the first cyclotron harmonic in cylindrical geometry were considered. It is a pity, but detailed comparison of the results presented here with those in [9] is impossible because of the following reasons. First, the authors of [9] used a pure numerical analysis. Second, they studied the surface O-mode with finite (non-zero) axial wavenumber and fixed azimuthal wavenumbers $m = \pm 1$. That is why their main result was the calculation of the phase velocity of the surface cyclotron waves as a function of axial wavenumber. Besides, the physical mechanisms of SCO-mode damping were not studied in [9].

**Fig. 5.2** SCO-mode eigenfrequency versus wavenumber for different $\beta_e$: $\beta_e = 0.5$ (1), $\beta_e = 1$ (2), $\beta_e = 10$ (3). $N_e = 10^3$

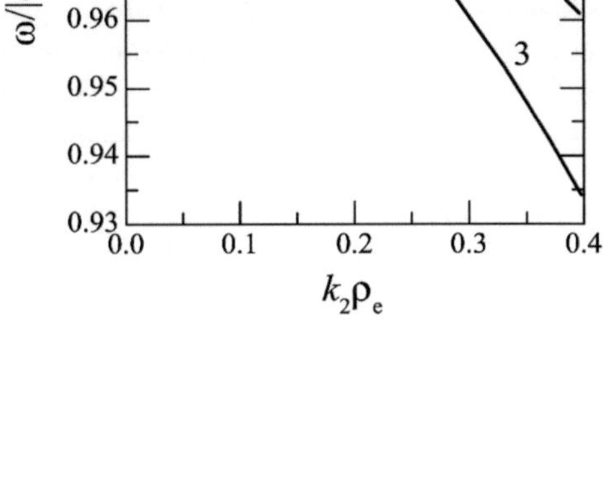

## 5.2 Damping of Surface Electron Cyclotron O-Modes

The physical mechanisms of electron SCO-mode damping are studied in this section. Corresponding damping rates are determined in non-relativistic approach for the case of uniform plasmas. Unlike the bulk cyclotron waves, corresponding surface waves experience also kinetic damping in addition to collisional damping. This kinetic damping is analogous to Landau damping for bulk waves. Its damping rate for SCO-modes is determined by the value $A_{33}$ (5.1). This channel of dissipation is called kinetic damping since its existence is associated with the interaction of the plasma particles with the plasma boundary interface when solving of the kinetic Vlasov–Boltzmann equation.

A large contribution to the total damping rate of SCO-modes also is due to plasma particle collisions. This channel is reasonably called as Ohmic damping, since the energy losses are proportional to the collision frequency , as in Ohm's law. It is very simple to take these losses into account by keeping a low collision frequency ($\nu \ll\omega$) in the component $\varepsilon_{33}(k_1, \omega, \nu)$ of the plasma dielectric permittivity tensor. To calculate the damping rate $\chi$ one has to replace $k_2$ by $k_2 + i\,\chi$, $|\chi| \ll |k_2|$ in the dispersion relation.

The total value of the damping rate $\chi$ consists of two terms: the damping rate caused by plasma particle collisions, $\chi_{\rm col}$, and that caused by electron interaction with the plasma interface, $\chi_{\rm kin}$.

In the limiting case of a thick dielectric layer, which separates the plasma from the waveguide metal wall, one can derive the following expression for the damping rate of electron SCO-modes caused by the plasma particle collisions:

$$\chi_{\rm col} \approx \nu k_2/[\varsigma\omega h_\alpha(s)], \tag{5.18}$$

where $\varsigma = 2$ in the case of the first and $\varsigma = 6$ for the second cyclotron harmonic, respectively. This result corresponds to the expressions (3.15) and (4.11) for the damping rates caused by the plasma particle collisions derived for the electron SCTM-modes and SCX-modes, respectively.

In the limiting case of a thin dielectric layer, the frequency shift $h_e(s)$ becomes proportional to the small factor $a_{\rm d}k_2 \ll 1$. That is why the damping rate caused by plasma particle collisions significantly increases with decreasing dielectric thickness: $\chi_{\rm col} \propto a_{\rm d}^{-2}$. Thus, the damping increases much compared with the case of thick dielectric layer. The damping rate of electron SCO-modes increases with decreasing thickness of the dielectric layer and with increasing number of the cyclotron harmonic (for fixed value of the product $k_2\rho_e$).

To calculate the kinetic damping rate $\chi_{\rm kin}$ of electron SCO-modes one has to take into account the anti-Hermitian part of the component $\varepsilon_{33}$ of the plasma permittivity tensor. Specifically, the term $A_{33}$ (5.1) determines the value of $\chi_{\rm kin}$. Asymptotic expressions of $A_{33}$ are inversely proportional to the square of the number of cyclotron harmonic. The same dependence is given also for electron SCX-modes [5, 54–57].

To derive the kinetic damping rate $\chi_{\rm kin}$ of electron SCO-modes one has to take into account the term $A_{33}$ in (5.12), and expand the other terms into a Taylor series in the same manner as it was done in Chap. 3 with respect to the linear term $i\chi_{\rm kin}\partial D_0/\partial k_2\big|_{\omega_0}$. In the case of electron SCO-modes at the first cyclotron harmonic, the expression for $\chi_{\rm kin}$ is given by:

$$|\chi_{\rm kin}| \approx \frac{8\rho_e}{3\sqrt{\pi}\,\delta k_2}\sqrt{\frac{\left(k_2^2\delta^2+1\right)h_e(1)}{\rho_e^2+4\delta^2 h_e(1)}}. \tag{5.19}$$

The following ratio is convenient to compare the kinetic damping rate with that caused by the plasma particle collisions in the case of a thick dielectric layer:

$$\frac{\chi_{\rm col}}{\chi_{\rm kin}} \approx \frac{\nu\delta^2}{\omega\rho_e^3 k_2}, \tag{5.20}$$

for electron SCO-modes at the first cyclotron harmonic. It follows from analysis of the ratio (5.20) that the kinetic damping can prevail over the collisional one ($\chi_{\rm kin} > \chi_{\rm col}$) in the case of a sufficiently hot plasma and weak external static magnetic field, when $\delta^3 < \rho_e^3$.

Reduction of the thickness of the protective dielectric coating, which prevents the direct contact of the plasma with the waveguide metal wall, results in a significant increase of the damping rate caused by the plasma particle collisions. That is why the SCO-mode collisional damping prevails over the kinetic one in the case of a thin dielectric coating ($a_{\rm d}k_2 \ll 1$), even in the case when the inequality $\delta^3 < \rho_e^3$ is valid. As an example, the condition $\chi_{\rm kin} < \chi_{\rm col}$ is valid for the following parameters of a technological device: $\nu = 10^5$ s$^{-1}$, $B_0 \approx 4 \times 10^4$ G and the plasma electrons should be sufficiently cold, $v_{Te}\sqrt{n_e} < 10^{15}$ s$^{-1}$ cm$^{-1/2}$ (here the thermal velocity is to be taken in cm/s, and the plasma particle density in cm$^{-3}$).

## 5.3 Influence of Finite Transverse Waveguide Dimensions and Inhomogeneity of Plasma Particle Density on Dispersion Properties of Surface Electron Cyclotron O-Modes

The plane waveguide wall, which is assumed to be made of metal with infinite conductivity, is placed in the region $x \leq -a_{\rm d}$. The infinite plasma occupies the half-space $x > 0$. The plasma particle density is inhomogeneous along the $\vec{x}$ axis (see Fig. 5.3). The plasma interface to a protective dielectric layer with dielectric constant $\varepsilon_{\rm d}$ and thickness $a_{\rm d}$ is located at the plane $x = 0$. The external static magnetic field $\vec{B}_0$ is oriented parallel to the plasma interface, along the $\vec{z}$ axis. The inhomogeneity of the plasma particle density is modelled by a set of uniform plasma layers of different width and particle density (see Fig. 5.3). This method for modelling the plasma particle density profile is well-known (see, e.g., [1, 10, 45]).

The set of Maxwell equations is used to describe the spatial field distribution of electron SCO-modes. Since the space is assumed to be uniform in the direction of the $\vec{y}$ axis, the dependence of SCO-mode fields on the coordinate $y$ and time $t$ is chosen in the following form: $E, H \propto \exp(ik_2 y - i\omega t)$, where $\omega$ and $\lambda = 2\pi/k_2$ are the angular eigenfrequency and wavelength, respectively. The case of sufficiently slow waves is under consideration, with $v_{\text{ph}}^2|\varepsilon_{ik}| \ll c^2$, where $c$ and $v_{\text{ph}} = \omega/k_2$ are the speed of light in vacuum and SCO-mode phase velocity, respectively, and the $\varepsilon_{ik}$ are the components of the plasma dielectric permittivity tensor. The space is assumed to be uniform along the $\vec{z}$ axis, so that $\partial/\partial z = 0$, and the spatial dispersion is assumed to be weak: $k_\perp v_e \ll \omega - s|\omega_e|$. Here $k_\perp^2 = k_1^2 + k_2^2$, $k_1$ is the component of the wave vector along the $\vec{x}$ axis, $v_e$ and $\omega_e$ are electron thermal velocity and cyclotron frequency, and $s$ is an integer, the number of cyclotron harmonic. The following considerations are restricted to the first harmonic of the electron cyclotron frequency.

The motion of plasma particle is described by the kinetic Vlasov–Boltzmann equation. The unperturbed plasma particle distribution function is chosen as Maxwellian. Interaction of plasma electrons with the plasma interface is taken into account. During solving the kinetic equation this interaction results in an anti-Hermitian part of the plasma permittivity tensor, which determines a specific damping of SCO-modes. This damping is analogous to Landau damping. That is why the relation between the density of RF currents in the plasma and the wave electric field has integral form, see [3] and a large number of references therein to original papers. These are devoted to studies on this problem in the cases of waves with different polarizations, for different models of particle reflection from the plasma interface, different ranges of wavelengths, etc. The papers [58–60] are of specific interest since they are devoted to modelling the magnetoactive plasma interface, calculating the wave fields, which penetrate into the plasma, determining wave absorption in the plasma, and searching

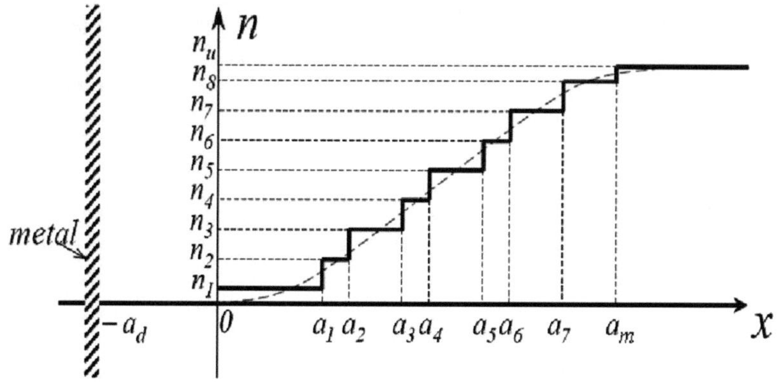


**Fig. 5.3** Schematic of the problem to be solved in the present Sect. 5.3

the solutions of the kinetic equation [61] for different models of particle reflection from the plasma interface. Approximate expressions for the non-difference parts of the kernels for the components of the plasma conductivity tensor were calculated in different limiting cases. For instance, damping of electron SCO-modes was studied in [52] in the limiting case of weak plasma spatial dispersion and for the model of plasma particle diffusive reflection from its interface. In this approach, the approximate expression (5.1) for the non-difference part $A_{33}$ of the kernel of the plasma conductivity tensor was calculated in [3].

Thus, if all the conditions listed above are valid, one can use the plasma permittivity tensor $\varepsilon_{ik}$ in the form which was obtained for the case of infinite magnetoactive plasma for studying the dispersion properties of SCO-modes (see, e.g., [39, 40]). In this case, the Fourier coefficients of axial electric current $j_3(k_\perp)$ and axial electric field $E_3(k_\perp)$ can be considered as linked with each other by the algebraic correlation (5.2), where $\sigma_{33}$ is the component of the infinite plasma conductivity tensor [39, 40] given by (5.3). Although the electron cyclotron resonance is under consideration, the contribution of ion terms to $\sigma_{33}$ should be estimated if $\omega \sim |\omega_e|$. Doing this, the ion temperature is assumed to be equal or less than the electron temperature. Keeping in mind the inequalities: $\Omega_e^2 \gg \Omega_i^2$, $k_\perp^2\,\rho_e^2 \ll k_\perp^2\,\rho_i^2 \ll 1$ and $\omega^2 - \omega_e^2 \ll \omega^2 - \omega_i^2 \sim \omega^2$, one can demonstrate that ion terms can be neglected and write the component $\sigma_{33}$ of the plasma conductivity tensor in the form (5.4).

The set of Maxwell equations decomposes into two independent subsets in this case. One describes the ordinarily polarized surface cyclotron wave with the field components $H_x$, $H_y$, $E_z$. Taking use of the Fourier method, one gets the following set of equations for the Fourier coefficients $H_1$, $H_2$, $E_3$ of the above mentioned field components:

$$\begin{cases} k_2 E_3 = k H_1; \\ k H_2 = i \partial E_z/\partial x; \\ \partial H_y/\partial x = i k_2 H_1 - i k \varepsilon_{33} E_3, \end{cases} \tag{5.21}$$

where $\varepsilon_{33} = 1 + 4\pi\, i\, \sigma_{33}/\omega$. Since the plasma is modelled as a set of layers with different values of plasma particle density and thickness, the set of equations (5.21) should be solved separately for every plasma region: $0 \le x \le a_1$; $a_1 \le x \le a_2$, etc. (here $a_n$ is the coordinate of the $n$-th layer). The SCO-mode fields should satisfy the following three boundary conditions. First, the tangential components of SCO-mode electric and magnetic fields should be continuous at the plasma–dielectric and plasma–plasma interfaces. Second, the tangential component of the SCO-mode electric field should be equal to zero at the waveguide metal wall. Third, SCO-mode fields should be finite in the waveguide, including the space $x \to \infty$. Application of these boundary conditions is typical for studies on surface wave propagation (see, e.g., [1, 43–45]), and makes it possible to derive the SCO-mode dispersion relation in this case.

The set of equations (5.21) can be used also for determination of the spatial distribution of the SCO-mode fields in the region of the dielectric layer which separates the plasma from the waveguide metal wall ($0 > x > -a_\mathrm{d}$), while keeping in mind the following replacement: $\varepsilon_{33} \rightarrow \varepsilon_\mathrm{d}$.

Here, this consideration is restricted to the case of the first electron cyclotron harmonic, although from the mathematical point of view, there is no principal problem to generalize the obtained results to the case of an arbitrary number of cyclotron harmonic $s$. Nevertheless, it is impossible to derive the dispersion relation in this general case.

Solving the set of equations (5.21) in the region of the dielectric layer makes it possible to derive the following expression for the dielectric impedance, $Z_\mathrm{d}(0) = E_x(0)/H_y(0)$, at its boundary with the plasma:

$$Z_\mathrm{d}(0) = -\tanh(|k_1|a_\mathrm{d}) i|k_1|/(k\varepsilon_\mathrm{d}). \tag{5.22}$$

Then, solving (5.21) for each plasma layer, one can derive the impedance, $Z_j \equiv E_z(a_j)/H_y(a_j)$, at both interfaces of the corresponding $j$-th plasma layer:

$$Z_0 = Z_\mathrm{d}(0);\; Z_j = R_j^2/\left(S_j - Z_{j-1}\right) - S_j; \quad \text{for } j = 1, 2, 3, \ldots, m. \tag{5.23}$$

In (5.23), $S_j = ik\left(q_j\xi_j\right)^{-1}\coth\left(q_j b_j\right)$, $R_j = ik\left[q_j\xi_j \sinh\left(q_j b_j\right)\right]^{-1}$, $q_j = k_2\sqrt{1 - k^2\varepsilon_\mathrm{pl}^{(j)} k_2^{-2}\xi_j^{-1}}$, $\xi_j = 1 + \left(\rho_e/\delta_e^{(j)}\right)^2\left[1 - (\omega_e/\omega)^2\right]^{-1}$, $q_j^{-1}$ is the depth of electromagnetic wave penetration into the $j$-th plasma layer, $\varepsilon_\mathrm{pl}^{(j)} = 1 - \Omega_e^2/\omega^2$ is the component of the cold plasma permittivity tensor, and $b_j = a_{j+1} - a_j$ is the thickness of the $j$-th plasma layer. For the region of uniform plasma ($x \geq a_m$), one can derive the following expression for the plasma impedance, $Z_u = ik/q_u\xi_u$. The subscript $u$ indicates that one deals with the impedance of a uniform plasma. Then the SCO-mode dispersion relation can be written in the following recurrent form:

$$Z_u = Z_m, \tag{5.24}$$

where $m$ is the total number of plasma layers. Although it is impossible to write down the expression for the impedance $Z_m$ for arbitrary values of $m$, it can be easily calculated in recurrent way on the base of formula (5.23) for the impedances at the interfaces of the different plasma layers, which are situated in front of the region of the uniform plasma.

A simple analytic expression (5.17) for the SCO-mode eigenfrequency at the first harmonic of the electron cyclotron frequency can be derived from the dispersion relation (5.24) in the limiting case of uniform plasma.

Numerical analysis of the dispersion relation (5.24) makes it possible to get the dispersion curves of electron SCO-modes for arbitrary values of the waveguide and plasma parameters. Comparison of the solutions of (5.24) for different number $m$ of plasma layers allows to conclude that the solutions practically coincide for $m > 25$ in the range $k_2\rho_e \leq 0.2$, and they differ from each other less than by 0.001 in the range $0.2 < k_2\rho_e \leq 0.5$. This confirms the applicability of this method of modelling the inhomogeneity of the plasma particle density for studying the dispersion properties of SCO-modes. That is why all the following numerical results are obtained for the case $m = 25$. The dependence of the SCO-mode eigenfrequency on the thickness $a_\mathrm{d}$ of the dielectric layer between the plasma and the metallic waveguide wall is presented in Fig. 5.4. The numerical analysis confirms that in absence of the dielectric layer, the tangential component of the SCO-mode electric field vanishes, as well as both magnetic components of the wave field. On the other hand, increase of the thickness of the layer stops to influence the SCO-mode dispersion properties when the thickness exceeds five Larmor radii. The dependence of the SCO-mode eigenfrequency on $\beta_e$ is presented in Fig. 5.5. It appears that it approaches to the corresponding cyclotron frequency if $\beta_e \to 0$. This is associated with the fact that the theoretical study of cyclotron waves needs careful account for the thermal motion of the plasma particles.

The results of studying the influence of the plasma particle density profile on the electron SCO-mode dispersion properties are presented in Fig. 5.6 for the cases

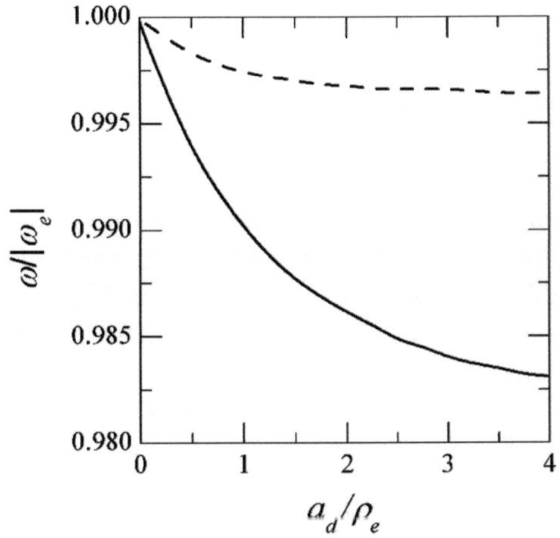


**Fig. 5.4** SCO-mode eigenfrequency versus the thickness of the dielectric layer for $k_2\rho_e = 0.3$, $\beta_e = 0.1$. Solid curve: $N_e = 10^3$, dashed curve: $N_e = 10^2$

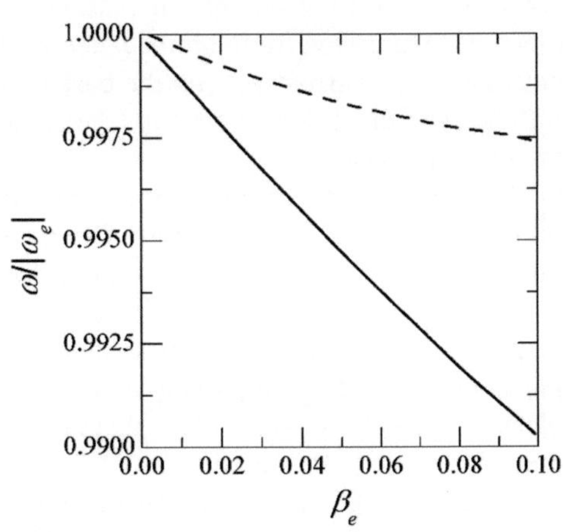


**Fig. 5.5** SCO-mode eigenfrequency versus the electron temperature for $k_2\rho_e = 0.3$, $a_\mathrm{d} = \rho_e$. Solid curve: $N_e = 10^3$, dashed curve: $N_e = 10^2$

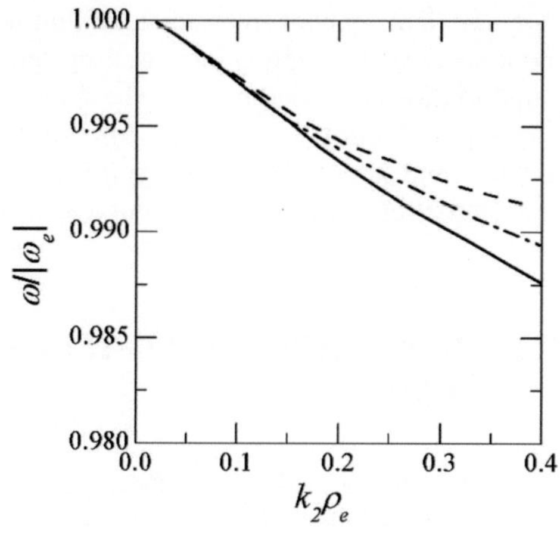


**Fig. 5.6** SCO-mode eigenfrequency versus the wavenumber in the case of inhomogeneous plasma for $N_e = 10^3$, $\beta_e = 0.1$, $a_d = \rho_e$. Dashed curve: $n = n_0[0.3 + 0.7(x/\rho_e)^{1/4}]$, dash-dotted curve: $n = n_0[0.8 + 0.2(x/\rho_e)]$, solid curve: $n = \text{const}$

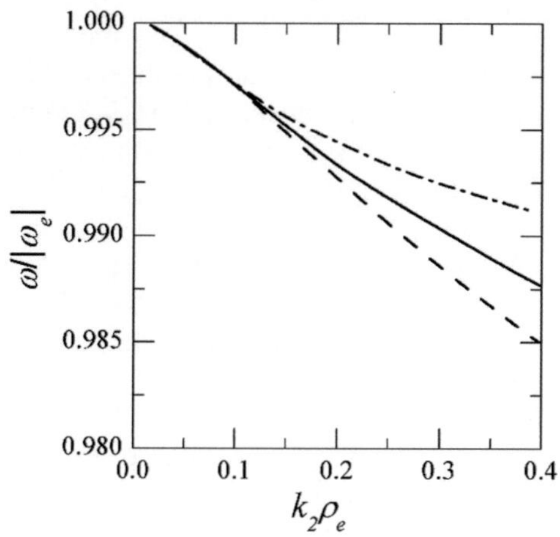


**Fig. 5.7** SCO-mode eigenfrequency versus the wavenumber in the case of inhomogeneous plasma with different signs of the gradients of the plasma particle density for $N_e = 10^3$, $\beta_e = 0.1$, $a_d = \rho_e$

of positive gradients of the plasma particle density, which means that the plasma particle density is minimal at the plasma interface and increases with increasing $x$ coordinate. Analysis of the presented results proves that a plasma particle density inhomogeneity weakly affects the SCO-mode dispersion properties. This influence is more pronounced in the range of short wavelengths.

Comparison of the electron SCO-mode eigenfrequencies for the cases of positive and negative gradients of the plasma particle density is presented in Fig. 5.7. Solid, dashed and dash-dotted curves correspond to the cases of uniform plasma, negative gradient (profile: $n = n_0\left[0.3 - 0.7(x/\rho_e)^{1/4}\right]$) and positive gradient (profile $n = n_0\left[0.3 + 0.7(x/\rho_e)^{1/4}\right]$) of the plasma particle density, respectively, for the same absolute value of the gradient. The values of electron SCO-mode eigenfrequencies calculated from (5.17) coincide with those calculated numerically in Fig. 5.6 for the case of uniform plasma particle density with the accuracy of $10^{-4}$.

Numerical analysis shows that accounting for the second component of the SCO-mode wave vector slightly increases the values of SCO-mode eigenfrequencies compared to those obtained in [52] where only one component of the wave vector was taken into account. The relative difference is about 2% in the range of long wavelengths, and about 7% in the region of short wavelengths. In general, the main characteristics of the dispersion curves are similar in these cases.

## 5.4 Beam Excitation of Surface Electron Cyclotron O-Modes

Surface cyclotron TM- and X-modes are known to actively interact with the flows of charged particles which move over the plasma interface [62–67]. To complete the physical picture of the interaction of surface cyclotron waves with different polarization with the flows of charged particles one has to complement the above investigations with the study of the possibility of beam excitation of the SCO-modes. In this section, the values of the SCO-mode growth rates are determined and the dependence of the growth rates on plasma and beam parameters is studied as well. The parameters of beam excitation of electron SCO-modes is compared with those of electron SCX-modes.

The geometry of the problem is as follows (see Fig. 5.8). A semibounded uniform plasma occupies the half-space $x \geq 0$. The flow of charged particles moves in the region $0 > x > -a$. The beam particle density is assumed to be much lower than that of the plasma. The metal wall of the waveguide is situated in the half-space $x \leq -a$. The external static magnetic field is oriented parallel to the plasma interface, along the $\vec{z}$ axis. The plasma spatial dispersion along the normal to the plasma interface is assumed to be weak, so that the following inequality is valid: $k_1^2 \rho_\alpha^2 \ll 2$. The space of the waveguide is assumed to be uniform along the $\vec{z}$ axis, so that $\partial/\partial z = 0$. The dependence of the surface wave fields on coordinates and time is chosen in the same form, as in Sect. 5.1, namely: $E_z$, $H_x$, $H_y \sim f(x) \exp(ik_2 y - i\omega t)$. The expression (5.23) for the component $\sigma_{33}$ of the plasma conductivity tensor is presented in the previous section. Remind that this expression is applicable in the approach of an infinite plasma. This means that the terms which account for the plasma particle interaction with its interface are neglected since they determine the wave damping, which is studied in Sect. 5.2.

Fourier transformation makes it possible to derive the following equation for the Fourier coefficient of the wave electric field:

$$E_3\left(k_2^2 - k^2\varepsilon_{33} + k_1^2\right) = i\,k\,\pi^{-1} H_y(x=0), \tag{5.25}$$

where $H_y$ $(x = 0)$ is the value of the magnetic wave field at the plasma boundary, and the expression (5.8) for the component $\varepsilon_{33}$ of the plasma permittivity tensor is presented in Sect. 5.1. The cases of the cyclotron resonance at the first and second

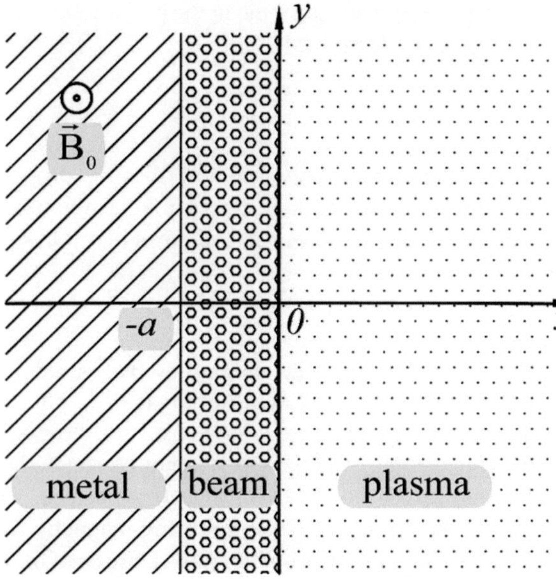


**Fig. 5.8** Schematic of the problem "beam excitation of SCO-modes"

harmonics are considered here. Then the plasma impedance at the interface $x = 0$ can be found from the (5.25) with the help of the theory of residues. To derive the dispersion relation, which describes the excitation of these surface waves, one has to calculate also the impedance of the region which is occupied by the beam.

The equations which describe the electromagnetic fields in the region $0 > x > -a$, where the charged particle beam moves with the velocity $\upsilon_0$, are similar to those (3.33) presented in Sect. 3.4 when one replaces the component $\varepsilon_{33}$ of the plasma permittivity tensor by the following expression:

$$\varepsilon_{\rm b} = 1 - \frac{\Omega_b^2}{\omega(\omega - k_2\upsilon_0)}, \tag{5.26}$$

where $\Omega_{\rm b}^2 = \theta \cdot \Omega_\alpha^2$, and $\theta = n_{\rm b}/n_{\rm pl}$. Since the beam particle density is assumed to be small compared to that of the plasma, the parameter $\theta$, which characterizes the correlation between the densities, is small: $\theta \ll 1$.

The impedance of the region occupied by the beam at the interface $x = 0$ has the following form:

$$\frac{E_z(0)}{H_y(0)} \approx \frac{ik}{k_2} \tanh(ak_2)[1 - D_{\rm b}], \quad D_{\rm b} = \frac{\omega \cdot \theta}{(\omega - k_2\upsilon_0)k_2^2\delta^2}. \tag{5.27}$$

To derive the dispersion relation which describes the initial stage of beam excitation of SCO-modes one has to choose the impedance (5.27) equal to the plasma impedance, whose value is calculated for the cases of SCO-modes at the first and second electron cyclotron harmonic in the previous section. Then the dispersion relation has the following form [68]:

$$D_0 = D_b, \tag{5.28}$$

where the function $D_0(\omega, k_2, a)$ describes the properties of the waveguide system without the beam. Its explicit expression has the most compact form in the case of SCO-modes at the first electron cyclotron harmonic:

$$D_0 \approx \left(1 + k_2^{-2}\delta^{-2}\right)\left[1 + \rho_e^2\delta^{-2}/4h_e(1)\right] \tanh^2(ak_2) - 1, \tag{5.29}$$

where $h_e(1)$ determines the shift of the SCO-mode eigenfrequency from the main cyclotron resonance, $\omega = s|\omega_e|(1 - h_e(s))^{-1}$. The explicit form of the expression for $D_0$ becomes much more complicated for the case of SCO-modes at the second cyclotron harmonic. That is why it is not presented here.

Since the SCO-modes are ordinarily polarized, which means that their electric field is oriented along the external static magnetic field, they can interact with the beam due to Cherenkov resonance only. Despite of the fact that these waves propagate at the harmonics of the cyclotron frequency, the conditions of Doppler resonance are not valid for them. This peculiarity of SCO-modes differs them from the case of SCX-modes [62]. The right hand side of the dispersion relation (5.28) has a sufficiently large magnitude under the conditions of weak beam particle density, $\theta \ll 1$, just in the case of Cherenkov resonance. This resonance condition can be written as follows: $\omega = k_2 v_0 + \gamma$, where the correction $\gamma$ is small, $|\gamma| \ll \omega$. Under these conditions, the term $D_b$ is inversely proportional to $\gamma$:

$$D_b \approx \omega\theta/\left(\gamma\, \delta^2\, k_2^2\right). \tag{5.30}$$

Searching for the maximum growth rate of resonant SCO-mode beam excitation at the first cyclotron harmonic which takes place under the resonance conditions, one obtains the following equation from (5.28):

$$\gamma(\gamma + i\nu_e)\left(1 + k_2^2\delta^2\right)\rho_e^2|\omega_e| = 4\,\theta\,\omega^3 h_e^2(1)\coth^2(ak_2), \tag{5.31}$$

where $\nu_e$ is the effective collision frequency of the plasma particles. It follows from the analysis of the left hand side of (5.31) that two modes of excitation of SCO-modes are possible: beam-plasma instability which takes place when $|\gamma| \gg \nu_e$, and dissipative beam instability which occurs if the opposite inequality is valid: $|\gamma| \ll \nu_e$. The growth rate of beam-plasma instability of SCO-modes at the first cyclotron harmonic can be written in the following form:

$$\mathrm{Im}\,(\gamma) \approx 2\sqrt{\theta}\,\omega\,\delta \frac{|h_e(1)\coth(ak_2)|}{\rho_e\sqrt{1 + k_2^2\,\delta^2}}. \tag{5.32}$$

In the case of the dissipative beam instability, the growth rate Im $(\gamma)$ has the other form:

$$\mathrm{Im}(\gamma) \approx 4\,\omega^2\,\theta \frac{\delta^2 h_e^2(1)\coth(ak_2)}{\nu_e \rho_e^2\left(1 + k_2^2\,\delta^2\right)}. \tag{5.33}$$

In the case of SCO-mode excitation at the second cyclotron harmonic, the growth rates have the following form under the conditions of beam-plasma and beam-dissipative instabilities, respectively:

$$\mathrm{Im}(\gamma) \approx \sqrt{\theta}\,\omega \frac{\left|8h_e^3(2)\right|^{1/4} k_2\,\delta^2}{\rho_e\left(1 + k_2^2\,\delta^2\right)^{1/4}} \coth(ak_2), \tag{5.34}$$

$$\mathrm{Im}(\gamma) \approx \frac{\theta \cdot \omega^2 k_2^2\,\delta^4 \sqrt{2h_e^3(2)}}{\nu_e\,\rho_e^2 \sqrt{1 + k_2^2\delta^2 \tanh^2(ak_2)}}. \tag{5.35}$$

Keeping in mind the significant dependence of the frequency shift $h_e(s)$ on the distance between the plasma and the metallic waveguide wall, one can conclude from analysis of the expressions (5.32)–(5.35) for the growth rate $\mathrm{Im}(\gamma)$, that a decrease of the product $ak_2$ results in decreasing growth rates of these instabilities. For instance, the shift for SCO-modes at the first cyclotron harmonic is given by:

$$h_e(1) \approx \frac{\rho_e^2\left(1 + k_2^2\,\delta^2\right)\sinh^2(ak_2)}{4\delta^2\left[k_2^2\,\delta^2 - \sinh(ak_2)\right]}. \tag{5.36}$$

That is why one can see from the (5.32) and (5.33) that in the range of long wavelengths ($ak_2 \ll 1$), the growth rates $\mathrm{Im}(\gamma)$ are proportional to $ak_2$ and $(ak_2)^3$ for the cases of beam-plasma and beam dissipative instabilities, respectively. Such dependence of the growth rate on the product $ak_2$ can be explained as follows. Reduction of the distance between the plasma interface and the waveguide metal wall worsens the conditions of SCO-mode existence: SCO-modes cannot propagate along the plasma–metal interface unlike the SCX-modes.

It was found for SCX-modes [62], that the value of the growth rate of beam excitation significantly depends on the transverse plasma dimension regardless of the influence of this dimension on the wave dispersion properties. That is why the present problem of the influence of the transverse plasma dimension on the efficiency of SCO-mode excitation is so important. The case of SCO-modes at the first cyclotron harmonic is considered for example. Account for the plasma finite transverse dimension $a_{\mathrm{pl}}$ changes the expression for the plasma impedance, and the expression for $D_0$ in (5.28) has to be written in the following form:

$$D_0 \approx \frac{k_2^2\,\delta^2 + 1}{k_2^2\,\delta^2}\left[1 + \frac{\rho_e^2}{4\,\delta^2 h_e(1)}\right] \frac{\tanh^2(ak_2)}{\tanh^2\left(a_{\mathrm{pl}}/\lambda_\perp\right)} - 1. \tag{5.37}$$

In (5.37), the depth $\lambda_\perp$ of SCO-mode penetration into the plasma in the case of the first electron cyclotron harmonic has the following form: $\lambda_\perp = \delta^2 k_2 \coth(ak_2)/\left(k_2^2\delta^2 + 1\right)$.

Comparison of the expression (5.37) with the corresponding expression (5.29) in the case of the model of semibounded plasma makes it possible to conclude that reduction of the plasma layer thickness results in a decreasing growth rate $\mathrm{Im}(\gamma)$ of the SCO-mode instability under beam excitation. That is why one can use the formulas (5.32)–(5.35) to calculate the growth rates of SCO-modes in the case of finite dimension of the plasma layer just by multiplying the formulas with $\tanh\left(a_{\mathrm{pl}}/\lambda_\perp\right)$. Growth rates of beam-plasma instability of SCO-modes are larger than those of beam dissipative instability as it should be according to the general theory of beam excitation of electromagnetic waves.

Finally, the values of SCO-mode growth rates are estimated. To calculate the SCO-mode growth rate $\mathrm{Im}(\gamma)$ in the case of beam-plasma instability at the first electron cyclotron harmonic one can use the following formula:

$$\mathrm{Im}(\gamma) \approx \sqrt{\theta}\,\Omega_e\, \upsilon_{Te}/(2c). \tag{5.38}$$

It follows from (5.38) that the growth rate $\mathrm{Im}(\gamma)$ increases with increasing plasma particle density and temperature. For the case of the plasma particle density $n_{\mathrm{pl}} = 10^{12}\ \mathrm{cm}^{-3}$, the expression (5.38) takes the more simple form: $\mathrm{Im}(\gamma) \approx \sqrt{\theta}\,\upsilon_{Te}\ \mathrm{s}^{-1}$ (here $\upsilon_{Te}$ is in cm/s). Comparison of the growth rates of SCO-modes, $\mathrm{Im}(\gamma)_{\mathrm{SCOM}}$, with those of SCX-modes, $\mathrm{Im}(\gamma)_{\mathrm{SCXM}}$, at the second cyclotron harmonic gives the following ratio [62]:

$$\frac{\mathrm{Im}(\gamma)_{\mathrm{SCOM}}}{\mathrm{Im}(\gamma)_{\mathrm{SCXM}}} \sim \frac{0.2}{\delta^3\, k_2^6\, \rho_e^3}. \tag{5.39}$$

That is why one can conclude that the growth rates of SCO-modes are larger than those of SCX-modes in the range of long wavelengths where $k_2^2\delta\rho_e < 1$.

In the present section SCO-mode excitation by flows of electrons which move over the plasma surface is investigated. This excitation takes place under the condition of Cherenkov resonance in the two scenarios of beam-plasma and beam-dissipative instabilities. The range of wavelengths has been derived, within which SCO-modes are excited more effectively than SCX-modes.

## 5.5 Gas Discharges Sustained by Surface Electron Cyclotron O-Modes

This section is devoted to studying the possibility to sustain a gas discharge by ordinarily polarized surface modes at harmonics of the electron cyclotron frequency (electron SCO-modes). A planar metal waveguide with protective dielectric coating of the metal walls plays the role of the discharge chamber. The external static magnetic field is oriented along the plasma–dielectric interface. SCO-modes are eigenmodes of this waveguide structure. Since generators which operate in the electron cyclotron frequency range are well developed, excitation of these modes does not look very complicated from the engineering point of view. Bulk cyclotron waves are well-known to be extensively used for various technological purposes [69, 70]. As far as the case of surface waves at the harmonics of the electron cyclotron frequency is concerned, their practical application is not so wide as compared with the case of the bulk waves, in particular, because of the lack of their theoretical investigation. Therefore, interest to studying the properties of SCO-modes is explained, in particular or first of all, by their possible application for sustaining such gas discharges. The first positive results concerning the possibility of using electron SCO-modes for sustaining gas discharges in external static magnetic fields, experimental study of the plasma spatial structure in such RF discharges, and experimental confirmation of the surface nature of the excited electromagnetic waves was presented in [71]. Direct measurements of the components of this wave electric field has confirmed that just the axial electric field component was dominant, and the radial field profile was typical for surface waves. This gave the reasons for the authors of [71] to claim that this wave was a SCO-mode. The dispersion properties of this wave with finite value of axial wavenumber and discrete values of azimuthal wavenumber were numerically studied in [9] for the model of a cylindrical plasma. Independently, the SCO-mode dispersion properties were studied by one of the authors of this monograph in [52] for the case of Cartesian geometry.

The electrodynamic model of a stationary gas discharge, which is sustained by SCO-modes, is constructed in this section. The dispersion properties of these SCO-modes are analytically studied in the Sects. 5.1 and 5.3 for the cases of a semibounded plasma and the plasma layer. The planar model of the discharge chamber considered here is as follows. The discharge plasma occupies the half-space $x \geq 0$. The protective dielectric coating with thickness $a_{\rm d}$ and dielectric constant $\varepsilon_{\rm d}$ is placed in the layer $-a_{\rm d} < x < 0$. The metal wall of the discharge chamber occupies the half-space $x \leq -a_{\rm d}$ (see Fig. 5.1). The external static magnetic field $\vec{B}_0$ is parallel to the plasma–dielectric interface and is oriented along the $\vec{z}$ axis. The electromagnetic fields of SCO-modes are described by the set of Maxwell equations, the dependence of the fields on coordinate and time is chosen in the following form: $E, H \sim \exp(ik_2 y - i\omega t)$, where $\omega$ and $k_2$ are the angular frequency and the component of the wave vector along the direction of SCO-mode propagation (along the $\vec{y}$ axis). Neither the waveguide parameters nor the wave properties depend on the $z$ coordinate. Under these conditions, the set of Maxwell equations can be separated into two independent subsets, where one describes SCO-modes with the components: $H_x$, $H_y$, $E_z$. The set of equations (5.6), which link the Fourier images of these fields, are presented in Sect. 5.1.

The motion of the plasma particles is described by the kinetic Vlasov–Boltzmann equation in the non-relativistic approach. Its solution is found by the method of characteristics. In the case of applying of the model of plasma particle diffusive reflection from the plasma surface, the link between the Fourier image of the axial component of the RF electric field in the plasma and the corresponding Fourier image of the SCO-mode electric field differs from that in the case of infinite plasma:

$$j_3(k_1) = \sigma_{33}(k_1)E_3(k_1) + I_3 E_3(k_1). \tag{5.40}$$

The following expression can be written for the component $\sigma_{33}(k_1)$ of the plasma conductivity tensor under the conditions of weak plasma spatial dispersion, $\left(|\vec{k}|\rho_e \ll 1\right)$, which is necessary for the further study of this model of the gas discharge:

$$\sigma_{33} \approx \frac{i}{4\pi} \sum_{\alpha} \sum_{n=-\infty}^{+\infty} \frac{\Omega_\alpha^2 (k_1 \rho_\alpha)^{2n}}{2^n n! \omega h_\alpha(n)}, \quad h_\alpha(n) = 1 - \frac{n\omega_\alpha}{\omega}. \tag{5.41}$$

Compared to the case of an infinite plasma, an additional term, which is proportional to $I_3$, arises in (5.40) just due to plasma particle collisions with the plasma surface. Under the conditions of weak plasma spatial dispersion, the expression for $I_3$ has the following form:

$$I_3 \approx \sum_{\alpha} \frac{\upsilon_{T\alpha}\, \Omega_\alpha^2}{\pi^{3/2} c\, \omega^2} \left[1 - \frac{1 + \exp(2i\pi\omega/\omega_\alpha)}{2 - 8\,\omega^2 \omega_\alpha^{-2}}\right]. \tag{5.42}$$

According to the conclusions of [4, 5] the difference between the $I_3$ values, which are calculated for different models of plasma particle interaction with the plasma surface, does not exceed 15% in the case of weak plasma spatial dispersion. That is why discussion of the problem, which interaction model is more appropriate for the considered case, is out of scope of this monograph.

Inverse Fourier transformation makes it possible to derive the following expression for the SCO-mode electric field in the plasma region:

$$E_z(x) = \int_{-\infty}^{+\infty} \frac{ik \exp(ik_1 x)}{\pi \Delta(k_1)} \left[H_y(0) + \pi I_3\right] \mathrm{d}k_1, \tag{5.43}$$

where $\Delta(k_1) = k_1^2 + k_2^2 - k^2 \varepsilon_{33}(k_1)$. The magnetic components of SCO-modes can be calculated if the expression (5.43) is known. The dispersion relation (5.24) for SCO-modes is derived in Sect. 5.3 in the non-relativistic approach, when $h_e(s) \gg \upsilon_{Te}^2 c^{-2}$. The case of electron SCO-modes at the first harmonic is considered below. Then the dispersion relation has the following form:

$$\left(k_2^2 - k^2 \varepsilon_3\right) \left(1 + \frac{\rho_e^2}{4\delta^2 h_e}\right) = k_2^2 \coth^2(k_2 a_\mathrm{d}), \tag{5.44}$$

where $\delta = c\, \Omega_e^{-1}$ is the skin-depth.

The transverse dimensions of the plasma, sustained by the discharge, can be estimated as the $1/e$-penetration depth of SCO-modes into the plasma. The dependence of the electric field of electron SCO-modes on the transverse coordinate in the plasma region can be found with the aid of the inverse Fourier transform of the set of equations (5.6):

$$E_z(x) = E_z(0)\ \tanh(k_2 a_\mathrm{d}) \exp\left(-x\sqrt{k_2^2 - k^2\varepsilon_\mathrm{pl}\phi^{-1}}/\sqrt{(\phi - k^2 k_2^{-2}\varepsilon_\mathrm{pl})\phi}\right), \tag{5.45}$$

where $\phi = 1 + \rho_e^2\,\omega^2(\omega^2 - \omega_e^2)\,\delta_e^{-2}$. In the region of vacuum, the expression for the SCO-mode electric field is as follows:

$$E_z(x) = E_z(0)\ \sinh[k_2(x + a_\mathrm{d})]\ \cosh^{-1}(k_2 a_\mathrm{d}). \tag{5.46}$$

The expressions (5.45) and (5.46) satisfy the boundary conditions: the tangential component of the SCO-mode electric field is continuous at the interface plasma–dielectric, and its amplitude decreases with going away from the plasma interface into the plasma core.

One can use the expressions (5.45) and (5.46) to calculate the value of the energy flux density carried by the SCO-mode, which is given by the Poynting vector $S_y = cE_z H_x/8\pi$. To do this, one needs also to know the normal component of the SCO-mode magnetic field. Thus, the following expression for the Poynting vector is obtained in the plasma region:

$$S_y(x) = \frac{-k_2\,\omega\,\exp\left(-2x\sqrt{k_2^2 - k^2\varepsilon_\mathrm{pl}\phi^{-1}}\right)}{8\pi\,\phi\,(k_2^2\,\phi - k^2\varepsilon_\mathrm{pl})}\tanh^2(k_2 a_\mathrm{d}) H_x^2(0). \tag{5.47}$$

In the region of vacuum, the Poynting vector is written in terms of hyperbolic functions:

$$S_y(x) = \frac{-\omega\,H_y^2(0)\ \sinh^2[k_2(x + a_\mathrm{d})]}{8\pi\,k_2\ \cosh^2[k_2 a_\mathrm{d}]}. \tag{5.48}$$

The dependencies (5.45)–(5.48) can be used for numerical analysis of the spatial distribution of the SCO-mode field and energy flux density for different plasma parameters. Figures 5.9 and 5.10 show the distributions for different values of plasma beta, $\beta_e = 4\pi\,n_e T_e/B_0^2$. One can see that an increase of $\beta_e$ causes an increasing depth of SCO-mode field penetration into the plasma.

Figures 5.11 and 5.12 present the distributions of electric field and Poynting vector for different values of the parameter $N_e = \Omega_e^2\,\omega_e^{-2}$ which characterizes the correlation between the plasma particle density and the external static magnetic field. Reduction of $N_e$ results in decreasing depth of the SCO-mode field penetration into the plasma.

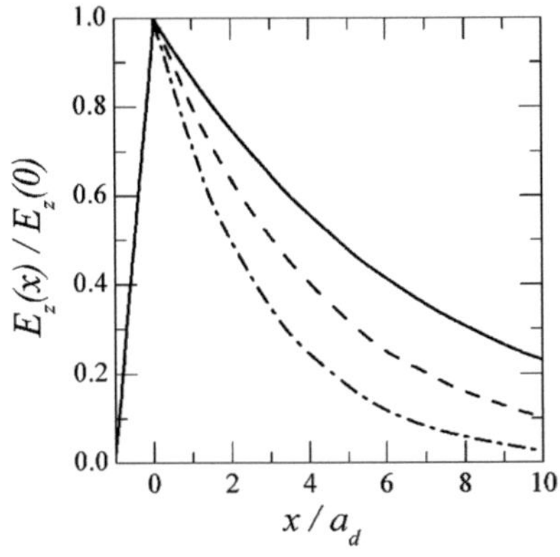


**Fig. 5.9** Spatial distribution of the SCO-mode electric field for different values of $\beta_e$. Solid curve: $\beta_e = 1.0$, dashed curve: $\beta_e = 0.5$, dash-dotted curve: $\beta_e = 0.1$

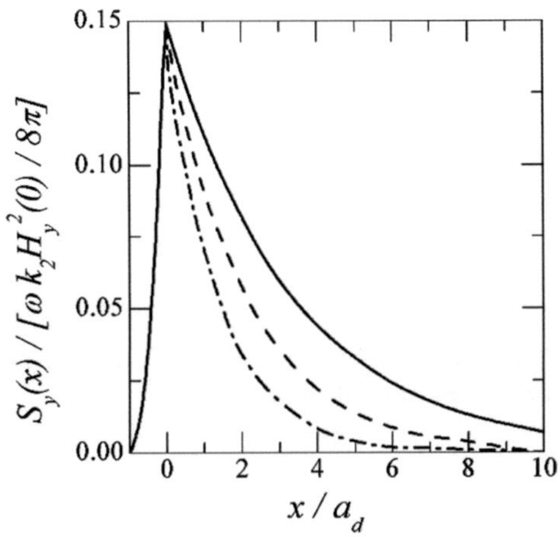


**Fig. 5.10** Spatial distribution of the energy flux density carried by the SCO-modes for different values of $\beta_e$. Solid curve: $\beta_e = 1.0$, dashed curve: $\beta_e = 0.5$, dash-dotted curve: $\beta_e = 0.1$

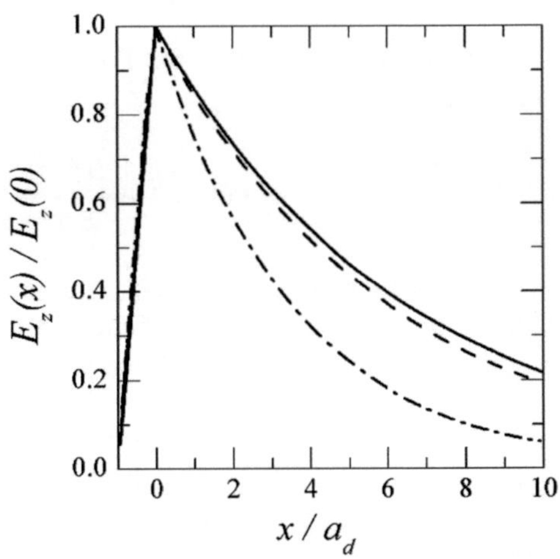


**Fig. 5.11** Spatial distribution of the SCO-mode electric field for different values of $N_e$. Solid curve: $N_e = 9.0$, dashed curve: $N_e = 4.0$, dash-dotted curve: $N_e = 1.0$

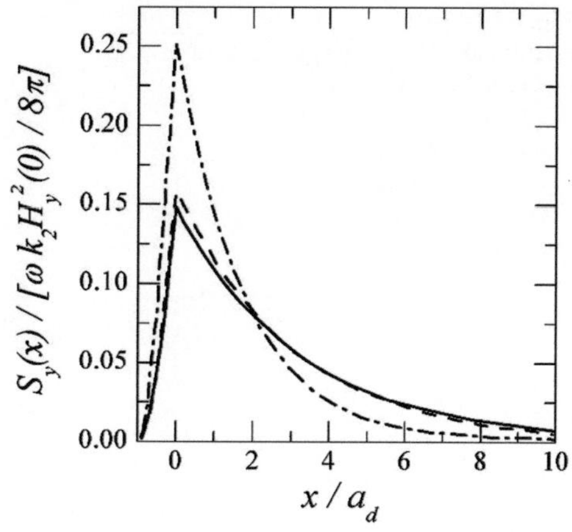


**Fig. 5.12** Spatial distribution of the energy flux density carried by the SCO-mode for different values of $N_e$. Solid curve: $N_e = 9.0$, dashed curve: $N_e = 4.0$, dash-dotted curve: $N_e = 1.0$

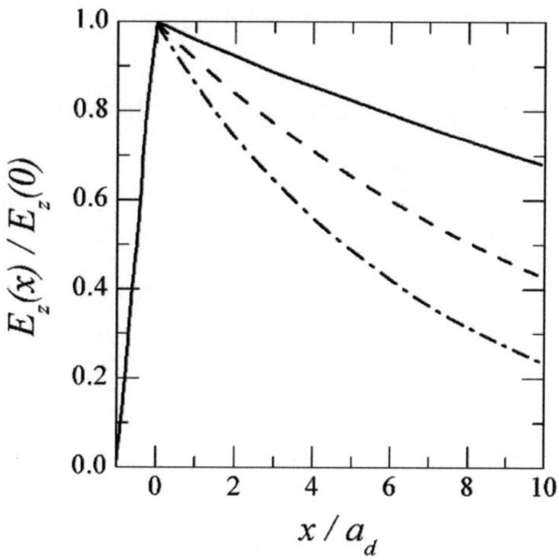


**Fig. 5.13** Spatial distribution of the SCO-mode electric field for different values of $k_2\rho_e$. Solid curve: $k_2\rho_e = 0.2$, dashed curve: $k_2\rho_e = 0.3$, dash-dotted curve: $k_2\rho_e = 0.4$

The spatial distributions of the SCO-mode electric field and the energy flux density strongly depend on the wavelength (parameter $k_2\,\rho_e$) as it is demonstrated in Figs. 5.13 and 5.14. The depth of wave penetration into the plasma increases with increasing wavelength.

The energy losses of electron SCO-modes during ionization of a neutral gas can be described by the energy balance equation, which can be written in the present case as follows:

$$dS_y/dy = -Q = -Q_{\text{col}} - Q_{\text{kin}}. \tag{5.49}$$

The energy of SCO-modes which is absorbed by the plasma per unit length of the gas discharge through the Ohmic, $Q_{\text{col}}$, and kinetic, $Q_{\text{kin}}$, channels are calculated below. The values $S_y$, $Q_{\text{col}}$, and $Q_{\text{kin}}$ in (5.49) are averaged over the time and transverse

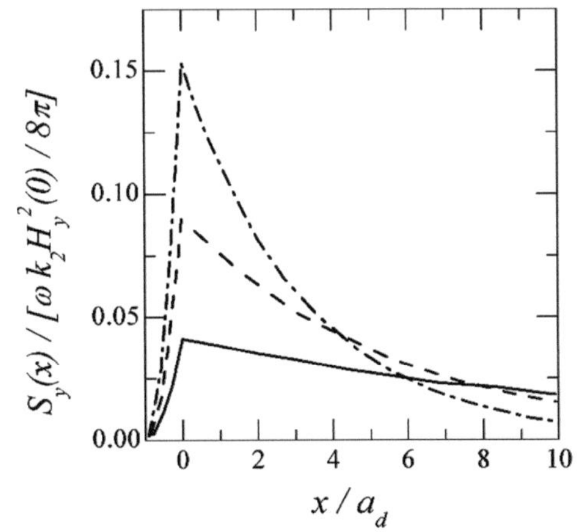


**Fig. 5.14** Spatial distribution of the energy flux density carried by the SCO-modes for different values of $k_2\rho_e$. Solid curve: $k_2\rho_e = 0.2$, dashed curve: $k_2\rho_e = 0.3$, dash-dotted curve: $k_2\rho_e = 0.4$

coordinate (in respect to the direction of wave propagation). These energies determine the efficiency of energy transfer from the SCO-modes to the plasma:

$$Q = Q_{\text{col}} + Q_{\text{kin}} = \frac{1}{2}\int_0^\infty j_z(x) E_z^*(x)\, \text{d}x. \tag{5.50}$$

If one considers plasma particle collisions only, one can derive an explicit expression for that part of energy which is absorbed through the Ohmic channel of dissipation:

$$Q_{\text{col}} \approx \frac{\nu_e\, k_2\, \delta^4 H_y^2(0)}{4\pi\, \rho_e^3 \left(1 + k_2^2\, \delta^2\right) \cosh^3(k_2 a_{\text{d}})\, \sinh(k_2 a_{\text{d}})}. \tag{5.51}$$

The value $Q_{\text{kin}}$ is determined by the plasma particle interaction with the interface of the gas discharge plasma:

$$Q_{\text{kin}} \approx \frac{v_{T\alpha}\, \Omega_\alpha^2\, H_y^2(0)}{6\pi^{3/2}\omega^2 \coth^2(k_2 a_{\text{d}})}. \tag{5.52}$$

Analysis of the expressions (5.51) and (5.52) for the SCO-mode energy absorbed by the plasma per unit length of the gas discharge through the different processes shows that the Ohmic channel dominates over the kinetic one, $Q_{\text{col}} > Q_{\text{kin}}$, in the case when the thickness of the dielectric coating is small, $k_2 a_{\text{d}} \ll 1$. The opposite situation, $Q_{\text{col}} < Q_{\text{kin}}$, takes place for the case of relatively thick dielectric coating, $k_2 a_{\text{d}} > 2$. This is explained by the strong dependence of the SCO-mode dispersion properties and the $Q$ value on the product $k_2 a_{\text{d}}$. That is why the right hand side of the energy balance equation (5.49) can be represented by one term (either $Q_{\text{col}}$ or $Q_{\text{kin}}$) in that or the other wavelength range.

The total SCO-mode energy flux

$$S_y = \frac{c}{8\pi} \int_{-a_d}^{+\infty} E_z H_x \, dx, \tag{5.53}$$

consists of the flux $S_d$, transported through the dielectric layer $-a_d < x < 0$, and the flux $S_{pl}$, transported through the plasma. The SCO-mode energy flux in the plasma region significantly depends on the thickness of the protective dielectric coating on the metal wall of the discharge chamber. This is explained by the fact that a reduction of the thickness of this coating results in worsening of the conditions for SCO-mode propagation, as it is demonstrated in Sect. 5.3. Substituting the expressions for these mode fields into the integral expression (5.53), one can derive the following expression for the SCO-mode energy flux $S_{pl}$ in the plasma:

$$S_{pl} = -\frac{\omega k_2}{2\pi} \int_0^{+\infty} \frac{h_e(1)\, H_y^2(0) \exp\left(-2x\lambda_\perp\right) dx}{\left(4\delta^2 h_e(1) + \rho_e^2\right)\left(k_2^2 \delta^2 + 1\right)} = -\frac{\omega H_y^2(0) \tanh(ak_2)}{16\pi \left(k_2^2 + \delta^{-2}\right)}. \tag{5.54}$$

The SCO-mode energy flux through the dielectric layer $(0 > x > -a_d)$ is defined as follows:

$$S_d = \frac{\omega H_y^2(0)\,(1 - \exp(-2a/\lambda_\perp))}{16\,\pi \left(k_2^2 + \delta^{-2}\right) \coth(ak_2)}. \tag{5.55}$$

Comparison of the expressions (5.54) and (5.55) for the SCO-mode energy fluxes makes it possible to conclude: in the range of long wavelengths ($a_d k_2 \ll 1$) $S_{pl} \sim S_d$, and in the range of short wavelengths ($a_d k_2 \gg 1$) most part of the SCO-mode energy propagates in the plasma region, $S_{pl} > S_d$:

$$S_y = -\frac{\omega\, H_y^2(0)}{16\,\pi\,\left(k_2^2 + \delta^{-2} \coth(ak_2)\right)}. \tag{5.56}$$

The other physical parameter, which determines the volume of the plasma sustained by the gas discharge, is the discharge length $L$. It can be derived from the solution of the energy balance equation (5.49) presented in the form (4.67) of the dependence of the plasma particle density on the $x$ coordinate. Under the conditions of SCO-mode damping through the Ohmic channel of dissipation, the discharge length is as follows:

$$L_{col} \approx \frac{\omega\, \rho_e^2\, \sinh^2(k_2 a_d)}{16\, \nu_e\, k_2\, \delta^2\, \varsigma}, \tag{5.57}$$

where the factor $\varsigma = 1$ in the range of sufficiently long waves ($k_2\,\delta \ll 1$), and $\varsigma = 2\,k_2^2\,\delta^2$ in the range of relatively short waves ($k_2\,\delta > 1$). In the case, when the kinetic mechanism of SCO-mode energy dissipation prevails over the Ohmic one, which is realized under low discharge pressure, the discharge length is given by the following expression:

$$L_{\mathrm{kin}} \approx \zeta \sqrt{\pi}\, \omega^2 \coth(a_{\mathrm{d}} k_2) \big(\rho_e\, c^2\big)^{-1}. \tag{5.58}$$

In (5.58), the factor $\zeta = 1.4\,\delta^4$ in the range of relatively long waves ($k_2\,\delta \ll 1$), and $\zeta = 0.4\,k_2^{-4}$ in the opposite range ($k_2\delta > 1$). One can conclude from the comparison of the expressions (5.57) and (5.58), that the discharge length rapidly decreases, proportionally to $\exp(4k_2 a_{\mathrm{d}})$, with decreasing thickness of the dielectric, under the conditions of high pressure of the working gas, when the Ohmic dissipation determines the energy transfer from SCO-modes to the plasma. One should keep in mind that $Q_{\mathrm{col}} > Q_{\mathrm{kin}}$ in the case of thin protective dielectric layers. The ratio $L_{\mathrm{kin}}/L_{\mathrm{col}}$ can be increased by enhancing the SCO-mode wavelength and/or external static magnetic field, as well as by decreasing the plasma temperature.

The obtained results make it possible to compare the parameters of RF discharges which can be sustained by SCTM-modes, SCX-modes, and SCO-modes. It appears that the discharge length is the largest in discharges sustained by SCX-modes, the smallest in the case of SCTM-modes, and has an intermediate value in the case of SCO-mode application.

As far as the transverse dimension of the discharge is concerned, the following conclusions can be made. The SCTM-modes penetrate most deeply into the plasma. Application of SCX-modes is characterized by smaller values of $\lambda_\perp$, which are either larger or of the same order as those which are achievable via application of SCO-modes for sustaining the gas discharge.

In the following, the parameters of a gas discharge sustained by SCO-modes are numerically estimated. For an RF generator with the working frequency 2.45 GHz (e.g. a magnetron), the external static magnetic field $B_0 = 857$ G, a 0.1 cm thick vacuum layer of the metallic discharge chamber, the produced plasma particle density of $n_{\mathrm{pl}} \approx 5 \times 10^{11}$ cm$^{-3}$, and the plasma temperature $T_e \approx 1$ eV, one gets: $\Omega_e^2 \approx 4.2\omega_e^2$, $\rho_e \approx 0.004$ cm. Under low pressure of the gas in the discharge, Ohmic heating becomes inefficient, and the kinetic absorption of the wave energy becomes the main channel of energy transfer. Then the discharge length in cm can be evaluated from the following formula: $L_{\mathrm{kin}} \approx 41.25\,k_2^{-4} \coth(k_2 a_{\mathrm{d}})$. In the case of SCO-modes with the wavelength of 1.5 cm, the discharge length can be calculated to be 50 cm, and the depth of wave penetration into the plasma as 4 cm [52, 72, 73]. Thus, orientation of the magnetic field relative to the plasma surface and choice of the working mode strongly influence the discharge parameters. It is reasonable to compare the obtained estimations with the results presented in [71], where a plasma with the particle density of $3 \times 10^{11}$ cm$^{-3}$ was experimentally obtained with axial and radial dimensions of the uniform plasma equal to 60 and 5.5 cm, respectively. This allows to state a good agreement of the theoretical calculations presented in [52, 72, 73] with the experimental measurements presented in [71].

Figure 5.15 shows the results of a numerical study of the discharge length. The discharge is sustained by the SCO-modes at the first harmonic of the electron cyclotron

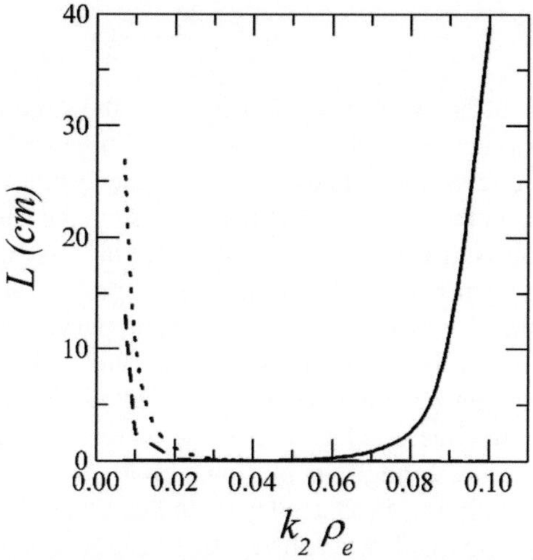


**Fig. 5.15** Length of the gas discharge sustained by SCO-modes versus wavenumber normalized by electron Larmor radius. Solid curve corresponds to the case of high pressure gas discharge (Ohmic damping of SCO-modes prevails over the kinetic one). Both dashed and dash-dotted curves correspond to the case of low pressure gas discharge (kinetic damping prevails over the Ohmic one) with the thickness of the vacuum layer of 0.2 and 0.1 cm, respectively

frequency. The working frequency of the RF generator is $f = 2.45$ GHz and the electron temperature is $T_e = 1$ eV. One can see that in the regime of Ohmic damping, the discharge length increases with increasing wavenumber, and in the case of the kinetic damping, this dependence is opposite.

The electrodynamic theory of gas discharges, which are sustained by SCO-modes, is developed in this section. Two channels of energy transfer from SCO-modes to the discharge plasma are taken into account: the Ohmic and kinetic ones. Average values of the SCO-mode energy flux and the power absorbed under both high and low pressure of the working gas are calculated. The dimensions of the uniform plasma, which can be sustained by SCO-modes under different discharge conditions, are defined. The spatial distribution of the SCO-mode fields is investigated. The dependencies of the penetration depth and the discharge length on the parameters of the discharge structure are studied. The properties of the discharge plasma, which is sustained by surface electron cyclotron waves of different polarization, are compared. Increasing external static magnetic field results in increasing discharge length and volume of plasmas produced in surface electron cyclotron wave application. The discharge length increases also with increasing SCO-mode wavelength.

## 5.6 Conclusions

Ordinarily polarized surface waves can propagate in the structures plasma–dielectric–metal at the harmonics of the electron cyclotron frequency. Change in the direction of the external static magnetic field does not influence the dispersion properties of these modes unlike in the case of SCX-modes. Like in the case of SCTM-modes, decrease of the thickness of the dielectric coating (or increase of the dielectric constant of this coating) results in worsening the conditions for SCO-mode existence: the rate of their damping, which is caused by plasma particle collisions, increases. For the case of finite transverse dimensions of plasma and dielectric layers, the SCO-mode kinetic damping rate is smaller than that caused by plasma particle collisions. If the dielectric layer thickness is larger than the wavelength, the main part of the wave energy is transported through the plasma region.

Electron beam excitation of SCO-modes is studied theoretically for both scenarios: resonant beam-plasma and beam-dissipative instabilities in the case when the electron beam flows above the plasma surface. Resonant excitation takes place under the condition of Cherenkov resonance. In the range of long wavelengths (compared to Larmor radius), the growth rates of SCO-modes are larger than those of SCX-modes.

The electrodynamic model of a stationary gas discharge, which is sustained by SCO-modes, is constructed in this chapter. The values of the energy fluxes, which are transported by SCO-modes in the case of a planar model of the discharge chamber and the external static magnetic field being parallel to the plasma working surface, are calculated. Analytical expressions for the SCO-mode power absorbed by the discharge plasma per unit length of the discharge are calculated for different cases of the working gas pressure. It is theoretically proved that under the conditions of low pressure of the gas discharge sustained by SCO-modes, the kinetic mechanism of wave energy transfer to the gas discharge plasma prevails over the Ohmic one.

The dimensions of the uniform plasma region, which can be sustained by application of SCO-modes, are calculated for different discharge conditions. Increase of the strength of the external static magnetic field is found to increase the discharge length and volume of the produced plasma. The discharge length increases as well with increasing SCO-mode wavelength.

## References

1. Aleksandrov, A. F., Bogdankevich, L. S., & Rukhadze, A. A. (1984). *Principles of plasma electrodynamics*. Berlin: Springer.
2. Kondratenko, A. N. (1976). *Plasma waveguides*. Moscow: Atomizdat. (in Russian).
3. Kondratenko, A. N. (1979). *Penetration of a field into plasmas*. Moscow: Atomizdat. (in Russian).
4. Kondratenko, A. N. (1985). *Surface and bulk waves in bounded plasmas*. Moscow: Energoatomizdat. (in Russian).
5. Beletski, N. N., Bulgakov, A. A., Khankina, S. I., & Iakovenko, V. M. (1984). *Plasma instabilities and non-linear phenomena in semi-conductors*. Kyiv: Naukova Dumka. (in Russian).

6. Vandenplas, P. E. (1968). *Electron waves and resonances in bounded plasmas*. London, N.Y.: Interscience Publishers.
7. Azarenkov, N. A., Kondratenko, A. N., & Ostrikov, K. N. (1993). Surface waves in plasma-metal structures. *Radiophysics and Quantum Electronics, 36*(5), 213–247.
8. Beletski, N. N., Svetlichnyj, V. M., Khalamejda, D. D., & Iakovenko, V. M. (1991). *Electromagnetic phenomena in HF range in inhomogeneous semi-conductor structures*. Kyiv: Naukova Dumka. (in Russian).
9. Ganguli, A., Akhtar, M. K., & Tarey, R. D. (1998). Theory of high-frequency guided waves in a plasma-loaded waveguide. *Physics of Plasmas, 5*(5), 1178–1189.
10. Kaner, E. A., & Skobov, V. G. (1967). Electromagnetic waves in metals in a magnetic field. *Soviet Physics Uspekhi, 9*(4), 480–503.
11. Vedenov, A. A. (1965). Solid state plasma. *Physics-Uspekhi, 7*(6), 809–822.
12. Lifshits, I. M., Azbel, M., & Kaganov, M. I. (1973). *Electron theory of metals*. New York: Consultants Bureau.
13. Steele, M., & Vural, B. (1969). *Wave interactions in solid state plasmas*. New York: McGraw Hill.
14. Vladimirov, V. V., Volkov, A. F., & Ye Meylikhov, Z. (1979). *Plasma of semiconductors*. Moscow: Atomizdat. (in Russian).
15. Dmitruk, N. L., Litovchenko, V. G., & Strizhevskiy, V. L. (1989). *Surface polaritons in semiconductors and dielectrics*. Kiev: Naukova Dumka. (in Russian).
16. Platzman, P. M., & Wolff, P. A. (1973). *Waves and interactions in solid state plasmas*. New York: Academic Press.
17. Bass, F. G., Bulgakov, A. A., & Tetervov, A. P. (1989). *High-frequency properties of semiconductors with superlattices*. Moscow: Nauka. (in Russian).
18. Nikitin, A. K., & Tishchenko, A. A. (1983). Surface electromagnetic waves and their application. *Zarubezhnaya elektronika, 3*, 38–55.
19. Wallis, R. F. (1972). Theory of surface magnetoplasmons in semiconductors. *Physical Review Letters, 28*(22), 1455–1458.
20. Bajbakov, V. I., & Datsko, V. N. (1981). Propagation of surface magneto-plasma waves in a semiconductor. *Fizika i tehnika poluprovodnikov, 15*(11), 2261–2263. (in Russian).
21. Brazis, R. S. (1981). Active and nonlinear interactions at the excitation of plasma-like polaritons in semiconductors. *Litovskiy Fizicheskiy Sbornik, 21*(4), 73–117. (in Russian).
22. Prozorovski, V. D., & Ocheret'ko, V. I. (1990). Determining the zonal and kinetic parameters of semiconductors with the help of magnetoplasma waves. *Zhurnal Tehnicheskoj Fiziki, 60*(2), 192–195. (in Russian).
23. Vukovic, S. (Ed.) (1986). *Surface waves in plasmas and solids*. Singapore: World Scientific.
24. Azarenkov, N. A., & Ostrikov, K. N. (1999). Surface magnetoplasma waves at the interface between a plasma-like medium and a metal in a Voigt geometry. *Physics Reports, 308*, 333–428.
25. Moisan, M., & Zakrzewski, Z. (1986). Plasmas sustained by surface waves at microwave and RF frequencies: Experimental investigation and applications. In J. M. Proud & L. H. Luessen (Eds.), *NATO Advanced Study Institute, Series B: Physics: Vol. 149. Radiative processes in discharge plasmas* (pp. 381–430).
26. Zhelyazkov, I., Atanasov, V., & Benova, E. (1986). Axial structure of a plasma column produced by a large-amplitude electromagnetic surface wave. In S. Vucovic (Ed.), *Surface waves in plasmas and solids* (pp. 467–489). Singapore: World Scientific.
27. Ferreira, C. M., & Moisan, M. (1993). Microwave discharges: Fundamentals and applications. In *NATO Advanced Study Institute, Series B: Physics: Vol. 302* (pp. 187–544).
28. Zhelyazkov, I., & Atanasov, V. (1995). Axial structure of low-pressure high-frequency discharges sustained by travelling electromagnetic surface waves. *Physics Reports, 255*, 79–201.
29. Nonaka, S. (1992). Mode identification of electromagnetic waves for large-area planar RF plasma productions. *Journal of the Physical Society of Japan, 61*(5), 1449–1452.
30. Nonaka, S. (1994). Proposal of non-cylindrical and large-area RF plasma production by surface waves. *Journal of the Physical Society of Japan, 63*, 3185–3186.

31. Nonaka, S. (1994). Very long and large-area RF plasma production by odd surface waves for online mass production of amorphous silicon solar cells or mirrors. *Japanese Journal of Applied Physics, 33,* 4226–4231.
32. Margot, J., & Moisan, M. (1992). Surface wave sustained plasmas in static magnetic fields for study of ECR discharge mechanisms. In M. Moisan & J. Pelletier (Eds.), *Microwave exited plasmas* (pp. 229–48). Amsterdam: Elsevier.
33. Moisan, M., & Zakrzewski, Z. (1986). Surface wave discharges in trapped tubes. In S. Vukovic (Ed.), *Surface waves in plasmas and solids* (pp. 605–608). Singapore: World Scientific.
34. Longinov, A. V., & Stepanov, K. N. (1992). Radio-frequency plasma heating in tokamaks in the ion-cyclotron frequency range. In A. G. Litvak (Ed.), *High-frequency plasma heating* (pp. 93–237). New York: AIP.
35. Becoulet, A. (1996). Heating and current drive regimes in the ion cyclotron range of frequency. *Plasma Physics and Controlled Fusion, 38*(12A), 1–12.
36. Laqua, H. P., Erckmann, V., Hartfuβ, H. J., et al. (1996). Resonant and non-resonant electron cyclotron heating at densities above the plasma cut-off by O–X–B mode conversion at W7-AS. In *23rd European Physical Society Conference on Controlled Fusion and Plasma Physics* (pp. 847–850), Ukraine.
37. Grigor'eva, L. I., Smerdov, B. I., Stepanov, K. N., et al. (1970). Plasma instability in a strong alternating electric field. *Soviet Physics JETP, 31*(1), 26–28.
38. Korzh, A. F., & Stepanov, K. N. (1988). Parametric instability of plasma in electric field of two pumping waves. *Fizika Plazmy, 14*(6), 698–705. (in Russian).
39. Akhiezer, A. I., Akhiezer, I. A., Polovin, R. V., Sitenko, A. G., & Stepanov, K. N. (1975). *Plasma electrodynamics*. Oxford: Pergamon Press.
40. Ginzburg, V. L., & Rukhadze, A. A. (1972). *Waves in magnetoplasma*. Heidelberg: Springer.
41. Lominadze, D. G. (1981). *Cyclotron waves in plasma*. Oxford: Pergamon Press.
42. Sitenko, O. G., & Mal'nev, V. M. (1994). *Principles of plasma theory*. Kyiv: Naukova Dumka. (in Ukrainian).
43. Gross, E. P. (1951). Plasma oscillations in a static magnetic field. *Physical Review, 82*(2), 232–242.
44. Sen, H. K. (1952). Solar "enhanced radiation" and plasma oscillations. *Physical Review, 88*(4), 816–822.
45. Sitenko, A. G., & Stepanov, K. N. (1957). On the oscillations of an electron plasma in a magnetic field. *Soviet Physics JETP, 4*(4), 512–520.
46. Bernstein, J. B. (1958). Waves in a plasma in a magnetic field. *Physical Review, 109*(1), 10–21.
47. Drummond, W. E., & Rosenbluth, M. N. (1962). Anomalous diffusion arising from microinstabilities in a plasma. *Physics of Fluids, 5*(12), 1507–1513.
48. Stix, T. H. (1990). Waves in plasmas: Highlight from the past and present. *Physics of Fluids B, 2*(8), 1729–1743.
49. Azbel, M. I., & Kaner, E. A. (1957). Theory of cyclotron resonance in metals. *Soviet Physics JETP, 5*(4), 730–744.
50. Kaner, E. A. (1958). Cyclotron resonance in plasma. *Soviet Physics JETP, 6*(2), 425–427.
51. Kaner, E. A. (1958). Theory of cyclotron resonance. *Journal of Experimental and Theoretical Physics, 6*(6), 1135–1138.
52. Girka, V. O. (2000). Propagation of the surface cyclotron O-modes at harmonics of ion and electron cyclotron frequencies. *Plasma Physics and Controlled Fusion, 42,* 999–1011.
53. Abramowitz, M., & Stegun, I. A. (1964). *Applied Mathematics, Series 55. Handbook of mathematical functions*. New York: National Bureau of Standards.
54. Girka, V. O., Girka, I. O., & Pavlenko, I. V. (1997). HF surface cyclotron waves in planar waveguide structures with nonuniform plasma filling. *Journal of Plasma Physics, 58*(1), 31–39.
55. Azarenkov, N. A., Girka, V. A., & Sporov, A. E. (1997). Ion cyclotron surface waves at the plasma-metal interface. *Plasma Physics Reports, 23*(3), 231–236.
56. Azarenkov, N. A., Girka, V. O., Kondratenko, A. M., et al. (1997). Surface waves on the harmonics of the electron cyclotron frequency propagating along a plasma–metal interface. *Plasma Physics and Controlled Fusion, 39,* 375–388.

57. Girka, V., Girka, I., Pavlenko, I., et al. (1996). Surface ion cyclotron propagation as a possible cause of plasma contamination during ion cyclotron resonance heating. In *Book of Abstracts 12-th International Conference on Plasma Surface Interactions in Controlled Fusion Devices* (pp. 199), France.
58. Dnestrovskii, Y. N., & Kostomarov, D. P. (1961). Electromagnetic waves in a half-space filled with a plasma. *Soviet Physics JETP, 12*(3), 587–591.
59. Arsenin, V. Y. (1963). On the absorption of an electromagnetic wave incident upon the plasma halfspace. *Izvestiya Vysshikh Uchebnykh Zavedenii Radiofizika, 6,* 457–460. (in Russian).
60. Bogomolov, Y. V. (1970). On the absorption of an electromagnetic wave in a bounded magnetically active plasma. *Radiophysics and Quantum Electronics, 13*(10), 1147–1151.
61. Meierovich, B. E. (1970). Influence of the sample surface on the cyclotron resonance in metals. *Soviet Physics JETP, 31*(1), 175–180.
62. Girka, V. O. (1998). Excitation of the surface type X-modes by charged particles beams. *Journal of Plasma Physics, 60*(2), 413–420.
63. Girka, V. O. (1984). *Surface cyclotron waves in plasmas.* Thesis for awarding candidate of sciences degree in Physics and Mathematics, Kharkiv State University, Ukraine. (in Russian).
64. Girka, V. A., & Kondratenko, A. N. (1982). Excitation of surface ion-cyclotron waves in a gyrotropic plasma by charged-particle beams. *Radiotekhnika i Elektronika, 27*(3), 534–539. (in Russian).
65. Girka, V. A., & Kondratenko, A. N. (1982). Excitation of surface cyclotron waves in a solid state plasma by charged-particle beams. *Zhurnal Tehnicheskoj Fiziki, 52*(8), 1521–1525. (in Russian).
66. Girka, V. A., & Peneva, J. K. (1982). Excitation of surface cyclotron waves in a plasma slab. *Ukrainskij Fizicheskij Zhurnal, 27*(8), 1246–1247. (in Russian).
67. Girka, V. O., Kondratenko, A. M., & Peneva, J. K. (1982). Surface cyclotron waves. In *Proceeding Conference on Surface Waves in Plasmas* (pp. 366–369), Bulgaria.
68. Girka, V. O. (2002). Excitation of surface cyclotron O modes by charged-particle beams. *Journal of Plasma Physics, 68*(2), 129–136.
69. Hiyamay, S., Onoz, T., Iizukaz, S., et al. (1996). Wide-area uniform plasma processing in an ECR plasma. *Plasma Sources Science and Technology, 5,* 299–304.
70. Miyazaway, W., Taday, S., Itoy, K., et al. (1996). A large-area ECR processing plasma. *Plasma Sources Science and Technology, 5,* 265–267.
71. Ganguli, A., Akhtar, M. K., & Tarey, R. D. (1999). Investigation of microwave plasmas produced in a mirror machine using ordinary-mode polarization. *Plasma Sources Science and Technology, 8,* 519–529.
72. Girka, V. O. (2001). Application of electron cyclotron surface modes to solids processing. *Physica Scripta, 63,* 484–490.
73. Girka, V. O., & Denysenko, I. B. (2000). Propagation of the surface cyclotron O-modes and their application for sustaining of microwave gas discharges. In *Proceedings of the International Congress on Plasma Physics ICPP-2000* (Vol. 1, pp. 9–12), Canada.

# Index




V. Girka et al., *Surface Electron Cyclotron Waves in Plasmas*,
Springer Series on Atomic, Optical, and Plasma Physics 107,
https://doi.org/10.1007/978-3-030-17115-5

MIX
Papier aus verantwortungsvollen Quellen
Paper from responsible sources
FSC® C105338

If you have any concerns about our products,
you can contact us on
**ProductSafety@springernature.com**

In case Publisher is established outside the EU,
the EU authorized representative is:
**Springer Nature Customer Service Center GmbH**
**Europaplatz 3, 69115 Heidelberg, Germany**

Printed by Libri Plureos GmbH
in Hamburg, Germany